Geostatistics for Environmental Scientists

Geostatistics for Environmental Scientists

Richard Webster
Rothamsted Experimental Station, UK

Margaret Oliver
University of Reading, UK

1339448

CHESTER COLLEGE

ACC. No.   DEPT.
01068770

CLASS No.
551·015195 WEB.

LIBRARY

JOHN WILEY & SONS, LTD

Chichester · New York · Weinheim · Brisbane · Singapore · Toronto

Copyright ©2001   John Wiley & Sons, Ltd
Baffins Lane, Chichester,
West Sussex, PO19 1UD, England

National   01243 779777
International (+44) 1243 779777

e-mail (for orders and customer service enquiries): cs-books@wiley.co.uk

Visit our Home Page on http://www.wiley.co.uk or http://www.wiley.com

All Rights Reserved. No part of this publication may be reproduced, stored in a retrieval system, or transmitted, in any form or by any means, electronic, mechanical, photocopying, recording, scanning or otherwise, except under the terms of the Copyright, Designs and Patents Act 1988 or under the terms of a licence issued by the Copyright Licensing Agency, 90 Tottenham Court Road, London W1P 9HE, UK, without the permission in writing of the Publisher and the copyright owner, with the exception of any material supplied specifically for the purpose of being entered and executed on a computer system, for the exclusive use by the purchaser of the publication.

Designations used by companies to distinguish their products are often claimed as trademarks. In all instances where John Wiley & Sons is aware of a claim, the product names appear in initial capital or all capital letters. Readers, however, should contact the appropriate companies for more complete information regarding trademarks and registration.

Other Wiley Editorial Offices

John Wiley & Sons, Inc., 605 Third Avenue,
New York, NY 10158-0012, USA

Wiley-VCH Verlag GmbH
Pappelallee 3, D-69469 Weinheim, Germany

Jacaranda Wiley Ltd, 33 Park Road, Milton,
Queensland 4064, Australia

John Wiley & Sons (Asia) Pte Ltd, 2 Clementi Loop #02-01,
Jin Xing Distripark, Singapore 129809

John Wiley & Sons (Canada) Ltd, 22 Worcester Road,
Rexdale, Ontario, M9W 1L1, Canada

Library of Congress Cataloging-in-Publication Data

Webster, R.
 Geostatistics for environmental scientists / Richard Webster, Margaret Oliver.
  p. cm—(Statistics in practice)
 Includes bibliographical references and index.
 ISBN 0-471-96553-7 (alk. paper)
  1. Geology—Statistical methods. I. Oliver, M. A. (Margaret A.) II. Title. III. Statistics in practice (Chichester, England)

QE33.2.S82.W43 2000
550'.7'2—dc21   00-031036

British Library Cataloguing in Publication Data

A catalogue record for this book is available from the British Library

ISBN 0-471-96553-7

Typeset in 10/12pt Lucida Bright by T&T Productions Ltd, London.
Printed and bound in Great Britain by Biddles Ltd, Guildford and King's Lynn
This book is printed on acid-free paper responsibly manufactured from sustainable forestry in which at least two trees are planted for each one used for paper production.

Contents

Appendix A   Aide-mémoire for spatial analysis   243

Appendix B   Genstat instructions for analysis   251

References   257

Index   265

Preface

When our *Statistical Methods in Soil and Land Resource Survey* was published ten years ago it was an instant success. It contained only two chapters on mainstream geostatistics. We had whetted readers' appetites, but readers and our students showed by their responses that they wanted to understand more and to have more detail of the techniques. We considered revising it, but realized that a completely new book was needed. That realization coincided with the editor's search for authors willing to write for 'Statistics in Practice', which we were pleased to do. This new book has had a long gestation as we have tested our presentation on newcomers to the subject in our taught courses and on practitioners with a modicum of experience. We have used the feedback to adjust its structure and content to guide the reader as thoroughly as we can in 271 pages.

Most of the material that we include is straightforward linear geostatistics using least-squares estimation. The theory and techniques have been around in mineral exploration and petroleum engineering for some three decades. For most of that time environmental scientists could not see the merits of the subject or appreciate how to apply it to their own problems, because of the context, the jargon and the mathematical presentation of the subject by many authors. This has changed dramatically in the last few years as soil scientists, hydrologists, ecologists, geographers and environmental engineers have seen that the technology is for them, if only they knew how to apply it. Here we have tried to satisfy that need.

The book is structured in the same way in which an environmental scientist would tackle an investigation. It begins with sampling, followed by data screening, summary and display. It considers some of the empirical methods that have been used for mapping, and then it introduces the theory of random processes, spatial covariances, and the variogram, which is central to practical geostatistics. Practitioners will learn how to estimate the variogram, what models they may use legitimately to describe it mathematically, and how to fit them. There is a brief excursion into the frequency domain to show the equivalence of covariance and spectral analysis. The book then returns to the principal reason for geostatistics, local estimation by kriging. Coregionalization is introduced as a means of improving estimates of a primary variable when data on one or more other variables are to hand or can be obtained readily, and the final chapter introduces disjunctive kriging, a non-linear method of prediction for decision making.

In environmental applications the problems are nearly always ones of estimation in two dimensions and of mapping. Rarely do they extend to three dimensions or are restricted to only one.

Geostatistics is not easy. No one coming new to the subject will read from cover to cover and remember everything that he or she should do. We have therefore added an aide-mémoire, which can be read and reread as often as may be. This will remind readers of what they should do and the order in which to do it. It is followed by some simple program instructions in the Genstat language for carrying out the analyses. These, with a few other commands to provide the necessary structures to read data and to write and display output, should enable practitioners to get started, after which they can elaborate their programs as their confidence and competence grow.

We illustrate the methods with data that we have explored previously in our research. The data are of soil properties, because we are soil scientists who use geostatistics in assessing soil resources. Nevertheless, there are close analogies in other branches of the environment at or near the land surface, which we have often had to include in our analyses and which readers will see as such.

The data come from surveys made by us or with our collaborators. The data for Broom's Barn Farm, accessible on the World Wide Web thanks to Dr J. D. Pidgeon, are from an original survey of the Farm soon after Rothamsted bought it in 1959, those for the Borders Region (Chapter 2) were collected by the Edinburgh School of Agriculture over some 20 years between 1960 and 1980 (provided by Mr R. B. Speirs). The data from the Jura used to illustrate coregionalization (Chapter 9) are from a survey made by the École Polytechnique Fédérale de Lausanne in 1992 under the direction of Mr J.-P. Dubois. Chapter 7 is based on a study of gilgai terrain in eastern Australia in 1973 by one of us when working with CSIRO (Australia), and the data from CEDAR Farm used to illustrate Chapter 9 were kindly provided by Dr Z. L. Frogbrook from her original study in 1998. We are grateful to the organizations and people whose data we have used.

The data from Broom's Barn Farm and the maps of it and of the simulated fields are available via the home page of the Statistics Department of Rothamsted Experimental Station at http://www.res.bbsrc.ac.uk/stats/.

Finally, we thank the publishers for allowing us to reproduce figures from our previous works. These include Blackwell Science Ltd (for Figures 5.14, 5.16 and 5.17), Oxford University Press (for Figures 5.5, 5.18, 5.20, 5.21 and 5.23), and Springer-Verlag (for Figures 5.6–5.11).

Richard Webster
Margaret Oliver
March 2000

Series Preface

Statistics in Practice is an important international series of texts which provide detailed coverage of statistical concepts, methods and worked case studies in specific fields of investigation and study.

With sound motivation and many worked practical examples, the books show in down-to-earth terms how to select and use an appropriate range of statistical techniques in a particular practical field within each title's special topic area.

The books meet the need for statistical support required by professionals and research workers across a range of employment fields and research environments. The series covers a variety of subject areas: in medicine and pharmaceutics (e.g. in laboratory testing or clinical trials analysis): in industry, finance and commerce (e.g. for design or forecasting): in the public services (e.g. in forensic science): in the earth and environmental sciences, and so on.

But the books in the series have an even wider relevance that this. Increasingly, statistical departments in universities and colleges are realizing the need to provide at least a proportion of their course-work in highly directed areas of study to equip their graduates for the work environment. For example, it is common for courses to be given on statistics applied to medicine, industry, social and administrative affairs, etc., and the books in this series provide support for such courses.

It is our aim to present judiciously chosen and well-written workbooks to meet everyday practical needs. Feedback of views from readers will be most valuable to monitor the success of this aim.

Vic Barnett
Series Editor
1999

1

Introduction

1.1 WHY GEOSTATISTICS?

Imagine the situation: a farmer has asked you to survey the soil of his farm. In particular, he wants you to determine the phosphorus content; but he will not be satisfied with the mean value for each field as he would have been a few years ago. He now wants more detail so that he can add fertilizer only where the soil is deficient, not everywhere. The survey involves taking numerous samples of soil, which you must transport to the laboratory for analysis. You dry the samples, crush them, sieve them, extract the phosphorus with some reagent and finally measure it in the extracts. The entire process is both time-consuming and costly. Nevertheless, at the end you have data from all the points from which you took the soil—just what the farmer wants, you might think!

The farmer's disappointment is evident, however. 'Oh', he says, 'this information is for a set of points, but I have to farm continuous tracts of land. I really want to know how much phosphorus the soil contains everywhere. I realize that that is impossible; nevertheless, I should really like some information at places between your sampling points. What can you tell me about those, and how do your small cores of soil relate to the blocks of land over which my machinery can spread fertilizer, that is, in bands 24 m wide?'

This raises further issues that you must now think about. Can you say what values to expect at intervening places between the sample points and over blocks the width of the farmer's fertilizer spreader? And how densely should you sample for such information to be reliable? At all times you must consider the balance between the cost of providing the information and the financial gains that will accrue to the farmer by differential fertilizing. In the wider context there may be an additional gain if you can help avoid over-fertilizing and thereby protect the environment from pollution by excess phosphorus. Your task, as a surveyor, is to be able to use sparse affordable data to estimate, or predict, the average values of phosphorus in the soil over blocks of land 24 m × 24 m or perhaps longer strips. Can you provide the farmer with spatially referenced values that he can use in his automated fertilizer spreader?

This is not fanciful. The technologically minded farmer can position his machines accurately to 2 m in the field, he can measure and record the yields of his crops continuously at harvest, he can modulate the amount of fertilizer he adds to match demand; but obtaining the information on the nutrient status of the soil at an affordable price remains a major challenge in modern precision farming (Lake *et al.*, 1997).

So, how can you achieve these aims? The answer is to use geostatistics— that is what it is for.

We can change the context to soil salinity, pollution by heavy metals, arsenic in ground water, rainfall, barometric pressure, to mention just a few of the many variables and materials that have been and are of interest to environmental scientists. What is common to them all is that the environment is continuous, but in general we can afford to measure properties at only a finite number of places. Elsewhere the best we can do is to estimate, or predict, in a spatial sense. This is the principal reason for geostatistics—it enables us to do so without bias and with minimum error. It allows us to deal with properties that vary in ways that are far from systematic and at all spatial scales.

We can take the matter a stage further. Alert farmers and land managers will pounce on the word 'error'. 'Your estimates are subject to error', they will say, 'in other words, they are more or less wrong. So there is a good chance that if we take your estimates at face value we shall fertilize or remediate where we need not, and waste money, because you have underestimated, and not fertilize or fail to remediate where we should.' The farmer will see that he might lose yield and profit if he applies too little fertilizer because you overestimate the nutrient content of the soil; the public health authority might take too relaxed an attitude if you underestimate the true value of a pollutant. 'What do you say to that?', they may say.

Geostatistics again has the answer. It can never provide complete information, of course, but, given the data, it can enable you to estimate the probabilities that true values exceed specified thresholds. This means that you can assess the farmer's risks of losing yield by doing nothing where the true values are less than the threshold or of wasting money by fertilizing where they exceed it.

Again, there are analogies in many fields. In some estimating the conditional probabilities of exceeding thresholds are as important as the estimates themselves because there are matters of law involved. Examples include limits on the arsenic content of drinking water (what is the probability that a limit is exceeded at an unsampled well?) and heavy metals in soil (what is the probability that there is more cadmium in the soil than the statutory maximum?).

1.1.1 Generalizing

The above is a realistic, if colourful, illustration of a quite general problem. The environment extends more or less continuously in two dimensions. Its properties have arisen as the result of the actions and interactions of many different processes and factors. Each process might itself operate on several scales simultaneously, in a non-linear way, and with local positive feedback. The environment which is the outcome of these processes varies from place to place with great complexity and at many spatial scales, from micrometres to hundreds of kilometres.

The major changes in the environment are obvious enough, especially when we can see them on aerial photographs and satellite imagery. Others are more subtle, and properties such as the temperature and chemical composition can rarely be seen at all, so that we must rely on measurement and the analysis of samples. By describing the variation at different spatial resolutions we can often gain insight into the processes and factors that cause or control it, and so predict in a spatial sense and manage resources.

As above, measurements are made on small volumes of material or areas a few centimetres to a few metres across, which we may regard as point samples, known technically as *supports*. In some instances the supports are enlarged by taking several small volumes of material and mixing them to produce bulked samples. In others several measurements might be made over larger areas and averaged rather than recording single measurements. Even so, these supports are generally very much smaller than the regions themselves and are separated from one another by distances several orders of magnitude larger than their own diameters. Nevertheless, they must represent the regions, preferably without bias.

An additional feature of the environment not mentioned so far is that at some scale the values of its properties are positively related—*autocorrelated*, to give the technical term. Places close to one another tend to have similar values, whereas ones that are farther apart differ more on average. Environmental scientists know this intuitively. Geostatistics expresses this intuitive knowledge quantitatively and then uses it for prediction. There is inevitably error in our estimates, but by quantifying the spatial autocorrelation at the scale of interest we can minimize the errors and estimate them too.

Further, as environmental protection agencies set maximum concentrations, thresholds, for noxious substances in the soil, atmosphere and water supply, we should also like to know the probabilities, given the data, that the true values exceed the thresholds at unsampled places. Farmers and graziers and their advisers are more often concerned with nutrients in the soil and the herbage it grows, and they may wish to know the probabilities of deficiency, i.e. the probabilities that true values are less than certain thresholds. With some elaboration of the basic approach geostatistics can also answer these questions.

The reader may ask in what way geostatistics differs from the classical methods that have been around since the 1930s; what is the effect of taking into account the spatial correlation? At their simplest the classical estimators, based on random sampling, are linear sums of data, all of which carry the same weight. If there is spatial correlation, then by stratifying we can estimate more precisely or sample more efficiently or both. If the strata are of different sizes then we might vary the weights attributable to their data in proportion. The means and their variances provided by the classical methods are regional, i.e. we obtain just one mean for any region of interest, and this is not very useful if we want local estimates. We can combine classical estimation with stratification provided by a classification, such as a map of soil types, and in that way obtain an estimate for each type of class separately. Then the weights for any one estimate would be equal for all sampling points in the class in question and zero in all others. This possibility of local estimation is described in Chapter 3. In linear geostatistics the predictions are also weighted sums of the data, but with variable weights determined by the strength of the spatial correlation and the configuration of the sampling points and the place to be estimated.

Geostatistical prediction differs from classical estimation in one other important respect; it relies on spatial models, whereas classical methods do not. In the latter, survey estimates are put on a probabilistic footing by the design of the sampling into which some element of randomization is built. This ensures unbiasedness, and provides estimates of error if the choice of sampling design is suitable. It requires no assumptions about the nature of the variable itself. Geostatistics, in contrast, assumes that the variable is random, that the actuality on the ground, in the sea or in the air, is the outcome of one or more random processes. The models on which predictions are based are of these random processes. They are not of the data, nor even of the actuality that we could observe completely if we had infinite time and patience. Newcomers to the subject usually find this puzzling; we hope that they will no longer do so when they have read Chapter 4, which is devoted to the subject. One consequence of the assumption is that sampling design is less important than in classical survey; we should avoid bias, but otherwise even coverage and sufficient sampling points are the main considerations.

The desire to predict was evident in weather forecasting and soil survey in the early twentieth century, to mention just two branches of environmental science. However, it was in mining and petroleum engineering that such a desire was matched by the financial incentive and resources for research and development. Miners wanted to estimate the amounts of metal in ore bodies and the thicknesses of coal seams, and petroleum engineers wanted to know the positions and volumes of reservoirs. It was these needs that constituted the force originally driving geostatistics because better predictions meant larger profits and smaller risks of loss. The solutions to the problems of spatial estimation are embodied in geostatistics and they are now used widely in many branches of science with spatial information. The origins of the subject have also given it its particular flavour and some of its characteristic terms, such 'nugget' and 'kriging'.

There are other reasons why we might want geostatistics. The main ones are description, explanation and control, and we deal with them briefly next.

1.1.2 Description

Data from classical surveys are typically summarized by means, medians, modes, variances, skewness, perhaps higher-order moments, and graphs of the cumulative frequency distribution and histograms and perhaps box-plots. We should summarize data from a geostatistical survey similarly. In addition, since geostatistics treats a set of spatial data as a sample from the realization of a random process, our summary must include the spatial correlation. This will usually be the experimental or sample variogram in which the variance is estimated at increasing intervals of distance and several directions. Alternatively, it may be the corresponding set of spatial covariances or autocorrelation coefficients. These terms are described later. We can display the estimated semivariances or covariances plotted against sample spacing as a graph. We may gain further insight into the nature of the variation at this stage by fitting models to reveal the principal features. A large part of this book is devoted to such description.

In addition, we must recognize that spatial positions of the sampling points matter; we should plot the sampling points on a map, sometimes known as a 'posting'. This will show the extent to which the sample fills the region of interest, any clustering (the cause of which should be sought), and any obvious mistakes in recording the positions such as reversed coordinates.

1.1.3 Interpretation

Having obtained the experimental variogram and fitted a model to it, we may wish to interpret them. The shape of the points in the experimental variogram can reveal much at this stage about the way that properties change with distance, and the adequacy of sampling. Variograms computed for different directions can show whether there is anisotropy and what form it takes. The variogram and estimates provide a basis for interpreting the causes of spatial variation and for identifying some of the controlling factors and processes. For example, Chappell and Oliver (1997) distinguished different processes of soil erosion from the spatial resolutions of the same soil properties in two regions. Burrough *et al.* (1985) detected early field drains in a field in the Netherlands, and Webster *et al.* (1994) attempted to distinguish sources of potentially toxic trace metals from their variograms in the Swiss Jura.

1.1.4   Control

The idea of controlling a process is often central in time-series analysis. In it there can be feedback such that the results of the analysis are used to change the process itself. In spatial analysis the concept of control is different. In many instances we are unlikely to be able to change the spatial characteristics of a process; they are given. But we may modify our response. Miners use the results of analysis to decide whether to send blocks of ore for processing if the estimated metal content is large enough, or to waste if not. They may also use the results to plan the siting of shafts and the expansion of mines. The modern precision farmer may use estimates from a spatial analysis to control his fertilizer spreader so that it delivers just the right amount at each point in a field.

1.2   A LITTLE HISTORY

Although mining provided the impetus for geostatistics in the 1960s, the ideas had arisen previously in other fields, more or less in isolation. The first record appears in a paper by Mercer and Hall (1911) who had examined the variation in the yields of crops in numerous small plots at Rothamsted. They showed how the plot-to-plot variance decreased as the size of plot increased up to some limit. 'Student', in his appendix to the paper, was even more percipient. He noticed that yields in adjacent plots were more similar than between others, and he proposed two sources of variation, one that was autocorrelated and the other that he thought was completely random. In total, this paper showed several fundamental features of modern geostatistics, namely spatial dependence, correlation range, the support effect and the nugget, all of which you will find in later chapters. Mercer and Hall's data provided numerous budding statisticians with material on which to practice, but the ideas had little impact in spatial analysis for two generations.

In 1919 R. A. Fisher began work at Rothamsted. He was concerned primarily to reveal and estimate responses of crops to agronomic practices and differences in the varieties. He recognized spatial variation in the field environment, but for the purposes of his experiments it was a nuisance. His solution to the problems it created was to design his experiments in such a way as to remove the effects of both short-range variation, by using large plots, and long-range variation, by blocking, and he developed his analysis of variance to estimate the effects. This was so successful that later agronomists came to regard spatial variation as of little consequence.

Within 10 years Fisher had revolutionized agricultural statistics to great advantage, and his book (Fisher, 1925) imparted much of his development of the subject. He might also be said to have hidden the spatial effects and therefore to have held back our appreciation of them. But two agronomists, Youden and Mehlich (1937), saw in the analysis of variance a tool for revealing and estimating spatial variation. Their contribution was to adapt Fisher's concepts so as to analyse the spatial scale of variation, to estimate the variation from different separating distances, and then to plan further sampling in the light of the knowledge gained. Perhaps they did not appreciate the significance of their research, for they published it in the house journal of their institute, where their paper lay dormant for many years. The technique had to be rediscovered not once but several times by, for example, Krumbein and Slack (1956) in geology, and Hammond *et al.* (1958) and Webster and Butler (1976) in soil science. We describe it in Chapter 5.

We next turn to Russia. In the 1930s A. N. Kolmogorov was studying turbulence in the air and the weather. He wanted to describe the variation and to predict. He recognized the complexity of the systems with which he was dealing and found a mathematical description beyond reach. Nowadays we call it chaos (Gleick, 1988). However, he also recognized spatial correlation, and he devised his 'structure function' to represent it. Further, he worked out how to use the function plus data to interpolate optimally, i.e. without bias and with minimum variance (Kolmogorov, 1941); see also Gandin (1965). Unfortunately, he was unable to use the method for want of a computer in those days. We now know Kolmogorov's structure function as the variogram and his technique for interpolation as kriging. We deal with them in Chapters 4 and 8, respectively.

The 1930s saw major advances in the theory of sampling, and most of the methods of design-based estimation that we use today were worked out then and later presented in standard texts such as Cochran's *Sampling Techniques*, of which the third edition (Cochran, 1977) is the most recent, and that by Yates, which appeared in its fourth edition as Yates (1981). Yates's (1948) investigation of systematic sampling introduced the semivariance into field survey. Von Neumann (1941) had by then already proposed a test for dependence in time series based on the mean squares of successive differences, which was later elaborated by Durbin and Watson (1950) to become the Durbin–Watson statistic. Neither of these leads was followed up in any concerted way for spatial analysis, however.

Matérn (1960), a Swedish forester, was also concerned with efficient sampling. He recognized the consequences of spatial correlation. He derived theoretically from random point processes several of the now familiar functions for describing spatial covariance, and he showed the effects of these on global estimates. He acknowledged that these were equivalent to Jowett's (1955) 'serial variation function', which we now know as the variogram, and mentioned in passing that Langsaetter (1926) had much earlier used the same way of expressing spatial variation in Swedish forest surveys.

The 1960s bring us back to mining, and to two men in particular. D. G. Krige, an engineer in the South African goldfields, had observed that he could improve his estimates of ore grades in mining blocks if he took into account the grades in neighbouring blocks. There was an autocorrelation, and he worked out empirically how to use it to advantage. It became practice in the gold mines. At the same time G. Matheron, a mathematician in the French mining schools, had the same concern to provide the best possible estimates of mineral grades from autocorrelated sample data. He derived solutions to the problem of estimation from the fundamental theory of random processes, which in the context he called the theory of regionalized variables. His doctoral thesis (Matheron, 1965) was a *tour de force.*

From mining, geostatistics has spread into several fields of application, first into petroleum engineering, and then into subjects as diverse as hydrogeology, meteorology, soil science, agriculture, fisheries, pollution, and environmental protection. There have been numerous developments in technique, but Matheron's thesis remains the theoretical basis of most present-day practice.

1.3   FINDING YOUR WAY

We are soil scientists, and the content of our book is inevitably coloured by our experience. Nevertheless, in choosing what to include we have been strongly influenced by the questions that our students, colleagues and associates have asked us and not just those techniques that we have found useful in our own research. We assume that our readers are numerate and familiar with mathematical notation, but not that they have studied mathematics to an advanced level or have more than a rudimentary understanding of statistics.

We have structured the book largely in the sequence that a practitioner would follow in a geostatistical project. We start by assuming that the data are already available. The first task is to summarize them, and Chapter 2 defines the basic statistical quantities such as mean, variance and skewness. It describes frequency distributions, the normal distribution and transformations to stabilize the variance. It also introduces the chi-square distribution for variances. Since sampling design is less important for geostatistical prediction than it is in classical estimation, we give it less emphasis than in our earlier *Statistical Methods* (Webster and Oliver, 1990). Nevertheless, the simpler designs for sampling in a two-dimensional space are described so that the parameters of the population in that space can be estimated without bias and with known variance and confidence. The basic formulae for the estimators, their variances and confidence limits are given.

From there the practitioner who knows that he or she will need to compute variograms or their equivalents, fit models to them, and then use the models to krige can go straight to Chapters 5, 6 and 8, respectively. Then, depending on the circumstances, the practitioner may go on to cokriging in which additional variables are brought into play (Chapter 9) or disjunctive kriging for estimating the probabilities of exceeding thresholds (Chapter 10).

Before that, however, newcomers to the subject are likely to have come across various methods of spatial interpolation already and to wonder whether these will serve their purpose. Chapter 3 describes briefly some of the more popular methods that have been proposed and are still used frequently for prediction, concentrating on those that can be represented as linear sums of data. It makes plain the shortcomings of these methods. Soil scientists are generally accustomed to soil classification, and they are shown how it can be combined with classical estimation for prediction. It has the merit of being the only means of statistical prediction offered by classical theory. The chapter also draws attention to its deficiencies, namely the quality of the classification and its inability to do more than predict at points and estimate for whole classes.

The need for a different approach from those described in Chapter 3, and the logic that underpins it, are explained in Chapter 4. A brief description of regionalized variable theory or the theory of spatial random processes upon which geostatistics is based is also part of the subject of this chapter.

Chapter 5 returns the reader to the real world in which he or she has to estimate the variogram from data. It gives the usual computing formula for the sample variogram, usually attributed to Matheron (1965), and a more robust alternative suggested by Cressie and Hawkins (1980). Confidence intervals on estimates are wider than many practitioners like to think, and the chapter shows that at least 100 or 150 sampling points are needed, distributed fairly evenly over the area. The sample variogram must then be modelled by choosing a mathematical function that seems to have the right form and then fitting it to the observed values. There is probably no more contentious topic in practical geostatistics than this. The common simple models are listed and illustrated in Chapter 6. The legitimate ones are few because a model variogram must be such that it cannot lead to negative variances. Greater complexity can be modelled by combining simple models. We recommend a combination of statistical fitting of plausible models by weighted least-squares approximation, graphing the results, and comparing them by statistical criteria.

For data that appear periodic the covariance analysis may be taken a step further by computing power spectra. This detour into the spectral domain is the topic of Chapter 7.

The reader is now ready for geostatistical prediction, i.e. kriging. Chapter 8 gives the equations and their solutions, and guides the reader in programming them. The equations show how the semivariances from the modelled variogram are used in geostatistical estimation (kriging). This chapter shows how the kriging weights depend on the variogram and the sampling configuration in relation to the target point or block, how in general only the nearest data carry significant weight, and the practical consequences that this has for the actual analysis.

Chapter 9 pursues two themes. It describes how to calculate and model the combined spatial variation in two or more variables simultaneously, and the use of the model to predict one of the variables from it, and others with which it is cross-correlated, by cokriging.

Finally, in Chapter 10 we deal with the most difficult subject in this book, disjunctive kriging. Its aim is to estimate the probabilities, given the data, that true values of a variable at unsampled places exceed specified thresholds.

In each chapter we have tried to provide sufficient theory to underpin the mechanics of the methods. We then give the formulae, from which you should be able to program the methods (except for the variogram modelling in Chapter 6). Then we illustrate the results of applying the methods from our own experience.

2

Basic statistics

Before focusing on the main topic of this book, geostatistics, we want to ensure that readers have a sound understanding of the basic quantitative methods for obtaining and summarizing information on the environment. There are two aspects to consider: one is the choice of variables and how they are measured; the other, and more important, is how to sample the environment. This chapter deals with these. Chapter 3 will then consider how such records can be used for estimation, prediction and mapping in a classical framework.

The environment varies from place to place in almost every aspect. There are infinitely many places at which we might record what it is like, but practically we can measure it at only a finite number by sampling. Equally, there are many properties by which we can describe the environment, and we must choose those that are relevant. Our choice might be based on prior knowledge of the most significant descriptors or from a preliminary analysis of data to hand.

2.1 MEASUREMENT AND SUMMARY

The simplest kind of environmental variable is binary, in which there are only two possible states, such as present or absent, wet or dry, calcareous or non-calcareous (rock or soil). They may be assigned the values 1 and 0, and they can be treated as quantitative or numerical data. Other features, such as classes of soil, soil wetness, stratigraphy, and ecological communities, may be recorded qualitatively. These qualitative characters can be of two types: unordered and ranked. The structure of the soil, for example, is an unordered variable and may be classified into blocky, granular, platy, etc. Soil wetness classes—dry, moist, wet—are ranked in that they can be placed in order of increasing wetness. In both cases the classes may be recorded numerically, but the records should not be treated as if they were measured in any sense. They can be converted to sets of binary variables, called 'indicators' in geostatistics (see Chapter 10), and can often be analysed by non-parametric statistical methods.

The most informative records are those for which the variables are measured fully quantitatively on continuous scales with equal intervals. Examples include the soil's thickness, its pH, the cadmium content of rock, and the proportion of land covered by vegetation. Some such scales have an absolute zero, whereas for others the zero is arbitrary. Temperature may be recorded in kelvin (absolute zero) or in degrees Celsius (arbitrary zero). Acidity can be measured by hydrogen ion concentration (with an absolute zero) or as its negative logarithm to base 10, pH, for which the zero is arbitrarily taken as $-\log_{10} 1$ (in moles per litre). In most instances we need not distinguish between them. Some properties are recorded as counts, e.g. the number of roots in a given volume of soil, the pollen grains of a given species in a sample from a deposit, the number of plants of a particular type in an area. Such records can be analysed by many of the methods used for continuous variables if treated with care.

Properties measured on continuous scales are amenable to all kinds of mathematical operation and to many kinds of statistical analysis. They are the ones that we concentrate on because they are the most informative, and they provide the most precise estimates and predictions. The same statistical treatment can often be applied to binary data, though because the scale is so coarse the results may be crude and inference from them uncertain. In some instances a continuous variable is deliberately converted to binary, or to an 'indicator' variable, by cutting its scale at some specific value, as described in Chapter 10.

Sometimes, environmental variables are recorded on coarse stepped scales in the field because refined measurement is too expensive. Examples include the percentage of stones in the soil, the root density, and the soil's strength. The steps in their scales are not necessarily equal in terms of measured values, but they are chosen as the best compromise between increments of equal practical significance and those with limits that can be detected consistently. These scales need to be treated with some caution for analysis, but they can often be treated as fully quantitative.

Some variables, such as colour hue and longitude, have circular scales. They may often be treated as linear where only a small part of each scale is used. It is a different matter when a whole circle or part of it is represented. This occurs with slope aspect and with orientations of stones in till. Special methods are needed to summarize and analyse such data (see Mardia and Jupp, 2000), and we shall not consider them in this book.

2.1.1 Notation

Another feature of environmental data is that they have spatial and temporal components as well as recorded values, which makes them unique or deterministic (we return to this point in Chapter 4). In representing the data we must distinguish measurement, location and time. For most classical statistical analyses location is irrelevant, but for geostatistics the location must be specified. We shall adhere to the following notation as far as possible throughout this text. Variables are denoted by italics; an upper-case Z for random variables and lower-case z for a realization, i.e. the actuality, and also for sample values of the realization. Spatial position, which may be in one, two or three dimensions, is denoted by bold $\mathbf{x}$. In most instances the space is two-dimensional, and so $\mathbf{x} = \{x_1, x_2\}$, signifying the vector of the two spatial coordinates. Thus $Z(\mathbf{x})$ means a random variable Z at place $\mathbf{x}$, and $z(\mathbf{x})$ is the actual value of Z at $\mathbf{x}$. In general, we shall use bold lower-case letters for vectors and bold capitals for matrices.

We shall use lower-case Greek letters for parameters of populations and either their Latin equivalents or place circumflexes (ˆ), commonly called 'hats' by statisticians, over the Greek for their estimates. For example, the standard deviation of a population will be denoted by σ and its estimate by s or $\hat{\sigma}$.

2.1.2 Representing variation

The environment varies in almost every aspect, and our first task is to describe that variation.

Frequency distribution: the histogram and box-plot

Any set of measurements may be divided into several classes, and we may count the number of individuals in each class. For a variable measured on a continuous scale we divide the measured range into classes of equal width and count the number of individuals falling into each. The resulting set of frequencies constitutes the frequency distribution, and its graph (with frequency on the ordinate and the variate values on the abscissa) is the *histogram*. Figures 2.1 and 2.4 are examples. The number of classes chosen depends on the number of individuals and the spread of values. In general, the fewer the individuals the fewer the classes needed or justified for representing them. Having equal class intervals ensures that the area under each bar is proportional to the frequency of the class. If the class intervals are not equal then the heights of the bars should be calculated so that the areas of the bars are proportional to the frequencies.

Another popular device for representing a frequency distribution is the box-plot. This is due to Tukey (1977). The plain 'box and whisker' diagram, like those in Figure 2.2, has a box enclosing the interquartile range, a line showing the median (see below), and 'whiskers' (lines) extending from the limits of the interquartile range to the extremes of the data, or to some other values such as the 90th percentiles.

Both the histogram and the box-plot enable us to picture the distribution, see how it lies about the mean or median, and identify extreme values.

Cumulative distribution

The cumulative distribution of a set of N observations is formed by ordering the measured values, z_i, $i = 1, 2, \ldots, N$, from the smallest to the largest, recording the order, say k, accumulating them, and then plotting k against z. The resulting graph represents the proportion of values less than z_k for all $k = 1, 2, \ldots, N$. The histogram can also be converted to a cumulative frequency diagram, though such a diagram is less informative because the data are grouped.

The methods of representing frequency distribution are illustrated in Figures 2.1–2.6.

2.1.3   The centre

Three quantities are used to represent the 'centre' or 'average' of a set of measurements. These are the mean, the median and the mode, and we deal with them in turn.

Mean

If we have a set of N observations, z_i, $i = 1, 2, \ldots, N$, then we can compute their arithmetic average, denoted by $\bar{z}$, as

$$\bar{z} = \frac{1}{N} \sum_{i=1}^{N} z_i, \qquad (2.1)$$

This, the mean, is the usual measure of central tendency.

The mean takes account of all of the observations, it can be treated algebraically, and the sample mean is an unbiased estimate of the population mean. For *capacity variables*, such as the phosphorus content in the topsoil of fields or daily rainfall at a weather station, means can be multiplied to obtain gross values for larger areas or longer periods. Similarly, the mean concentration of a pollutant metal in the soil can be multiplied by the mass of soil to obtain a total load in a field or catchment. Further, addition or physical mixing should give the same result as averaging.

Intensity variables are somewhat different. These are quantities such as barometric pressure and matric suction of the soil. Adding them or multiplying them does not make sense, but the average is still valuable as a measure of the centre. Physical mixing will in general not produce the arithmetic average. Some properties of the environment are not stable in the sense that bodies of material react with one another if they are mixed.

For example, the average pH of a large volume of soil or lake water after mixing will not be the same as the average of the separate bodies of the soil or water that you measured previously. Chemical equilibration takes place. The same can be true for other exchangeable ions. So again, the average of a set of measurements is unlikely to be the same as a single measurement on a mixture.

Median

The median is the middle value of a set of data when the observations are ranked from smallest to largest. There are as many values less than the median as there are greater than it. If a property has been recorded on a coarse scale then the median is a rough estimate of the true centre. Its principal advantage is that it unaffected by extreme values, i.e. it is insensitive to outliers, mistaken records, faulty measurements and exceptional individuals. It is a robust summary statistic.

Mode

The mode is the most typical value. It implies that the frequency distribution has a single peak. It is often difficult to determine the numerical value. If in constructing a histogram the class interval is small then the mid-value of the most frequent class may be taken as the mode. For a symmetric distribution the mode, the mean and the median are in principle the same. For an asymmetric one

$$(\text{mode} - \text{median}) \approx 2 \times (\text{median} - \text{mean}). \qquad (2.2)$$

In asymmetric distributions the median and mode lie further from the longer tail of the distribution than the mean, and the median lies between the mode and the mean—see Figures 2.1(a) and 2.4(a).

2.1.4 Dispersion

There are several measures for describing the spread of a set of measurements: the range, interquartile range, mean deviation, standard deviation and its square, the variance. These last two are so much easier to treat mathematically, and so much more useful therefore, that we concentrate on them almost to the exclusion of the others.

Variance and standard deviation

The variance of a set of values, which we denote S^2, is by definition

$$S^2 = \frac{1}{N} \sum_{i=1}^{N} (z_i - \bar{z})^2. \qquad (2.3)$$

The variance is the second moment about the mean. Like the mean, it is based on all of the observations, it can be treated algebraically, and it is little affected by sampling fluctuations. It is both additive and positive. Its analysis and use are backed by a huge body of theory. Its square root is the standard deviation, S. Below we shall replace the divisor N by $N-1$ so that we can use the variance of a sample to estimate σ^2, the population variance, without bias.

Coefficient of variation

The standard deviation expresses dispersion in the same units as those in which the variable is measured. There are situations in which we may want to express it in relative terms, as where a property has been measured in two different regions to give two similar values of S but where the means are different. If the variances are the same we might regard the region with the smaller mean as more variable than the other in relative terms. The coefficient of variation (CV) can express this. It is usually presented as a percentage:

$$CV = 100(S/\bar{z})\%. \tag{2.4}$$

It is useful for comparing the variation of different sets of observations of the same property. It has little merit for properties with scales having arbitrary zeros and for comparing different properties except where they can be measured on the same scale.

Skewness

The skewness measures the asymmetry of the observations. It is defined formally from the third moment about the mean:

$$m_3 = \frac{1}{N} \sum_{i=1}^{N} (z_i - \bar{z})^3. \tag{2.5}$$

The coefficient of skewness is then

$$g_1 = \frac{m_3}{m_2\sqrt{m_2}} = \frac{m_3}{S^3}, \tag{2.6}$$

where m_2 is the variance. Symmetric distributions have $g_1 = 0$. Skewness is the most common departure from normality (see below) in measured environmental data. If the data are skewed then there is some doubt as to which measure of centre to use. Comparisons between the means of different sets of observations are especially unreliable because the variances can differ substantially from one set to another.

Kurtosis

The kurtosis expresses the peakedness of a distribution. It is obtained from the fourth moment about the mean:

$$m_4 = \frac{1}{N} \sum_{i=1}^{N} (z_i - \bar{z})^4. \tag{2.7}$$

The coefficient of kurtosis is given by

$$g_2 = \frac{m_4}{m_2^2} - 3 = \frac{m_4}{(S^2)^2} - 3. \tag{2.8}$$

Its significance relates mainly to the normal distribution, for which $g_2 = 0$. Distributions that are more peaked than normal have $g_2 > 0$; flatter ones have $g_2 < 0$.

2.2 THE NORMAL DISTRIBUTION

The normal distribution is central to statistical theory. It has been found to describe remarkably well the errors of observation in physics. Many environmental variables, such as of the soil, are distributed in a way that approximates the normal distribution. The form of the distribution was discovered independently by De Moivre, Laplace and Gauss, but Gauss seems generally to take the credit for it, and the distribution is often called 'Gaussian'. It is defined for a continuous random variable Z in terms of the probability density function (pdf), $f(z)$, as

$$f(z) = \frac{1}{\sigma\sqrt{2\pi}} \exp\left\{-\frac{(z-\mu)^2}{2\sigma^2}\right\}, \tag{2.9}$$

where μ is the mean of the distribution and σ^2 is the variance.

The shape of the normal distribution is a vertical cross-section through a bell. It is continuous and symmetrical, with its peak at the mean of the distribution. It has two points of inflexion, one on each side of the mean at a distance σ. The ordinate $f(z)$ at any given value of z is the *probability density* at z. The total area under the curve is 1, the total probability of the distribution. The area under any portion of the curve, say between z_1 and z_2, represents the proportion of the distribution lying in that range. For instance, slightly more than two-thirds of the distribution lies within one standard deviation of the mean, i.e. between $\mu - \sigma$ and $\mu + \sigma$; about 95% lies in the range $\mu - 2\sigma$ to $\mu + 2\sigma$; and 99.74% lies within three standard deviations of the mean.

Just as the frequency distribution can be represented as a cumulative distribution, so too can the pdf. In this representation the normal distribution is characteristically sigmoid as in Figures 2.3(a), 2.3(c), 2.6(a) and

2.6(c). The main use of the cumulative distribution function is that the probability of a value's being less than a specified amount can be read from it. We shall return to this in Chapter 10.

In many instances distributions are far from normal, and these departures from normality give rise to unstable estimates and make inference and interpretation less certain than they might otherwise be. As above, we can be in some doubt as to which measure of centre to take if data are skewed. Perhaps more seriously, statistical comparisons between means of observations are unreliable if the variable is skewed because the variances are likely to differ substantially from one set to another.

2.3   COVARIANCE AND CORRELATION

When we have two variables, z_1 and z_2, we may have to consider their joint dispersion. We can express this by their *covariance*, $C_{1,2}$, which for a finite set of observations is

$$C_{1,2} = \frac{1}{N} \sum_{i=1}^{N} \{(z_1 - \bar{z}_1)(z_2 - \bar{z}_2)\}, \tag{2.10}$$

in which $\bar{z}_2$ and $\bar{z}_2$ are the means of the two variables. This expression is analogous to the variance of a finite set of observations, equation (2.3).

The covariance is affected by the scales on which the properties have been measured. This makes comparisons between different pairs of variables and sets of observations difficult unless measurements are on the same scale. Therefore, the *Pearson product-moment correlation coefficient*, or simply the correlation coefficient, is often preferred. It refers specifically to linear correlation and it is a dimensionless value.

The correlation coefficient is obtained from the covariance by

$$r = \frac{C_{1,2}}{S_1 S_2}. \tag{2.11}$$

This quantity is a measure of the relation between two variables; it can range between 1 and -1. If units with large values of one variable also have large values of the other then the two variables are positively correlated, $r > 0$; if the large values of the one are matched by small values of the other then the two are negatively correlated, $r < 0$. If $r = 0$ then there is no linear relation.

Just as the normal distribution is of special interest for a single variable, for two variables we are interested in a joint distribution that is bivariate normal. The joint pdf for such a distribution is given by

$$f(z) = \frac{1}{2\pi\sigma_1\sigma_2\sqrt{1-\rho^2}} \exp\left[-\left\{\frac{(z_1-\mu_1)^2}{\sigma_1^2}\right.\right.$$
$$\left.\left. -\frac{2\rho(z_1-\mu_1)(z_2-\mu_2)}{\sigma_1\sigma_2} + \frac{(z_2-\mu_2)^2}{\sigma_2^2}\right\}\middle/2(1-\rho^2)\right]. \quad (2.12)$$

In this equation μ_1 and μ_2 are the means of z_1 and z_2, σ_1^2 and σ_2^2 are the variances, and ρ is the correlation coefficient.

One can imagine the function as a bell shape standing above a plane defined by z_1 and z_2 with its peak above the point $\{\mu_1,\mu_2\}$. Any vertical cross-section through it appears as a normal curve, and any horizontal section is an ellipse—a 'contour' of equal probability.

2.4   TRANSFORMATIONS

To overcome the difficulties arising from departures from normality we can attempt to transform the measured values to a new scale on which the distribution is more nearly normal. We should then do all further analysis on the transformed data, and if necessary transform the results to the original scale at the end. The following are some of the commonly used transformations for measured data.

2.4.1   Logarithmic transformation

The geometric mean of a set of data is

$$\overline{g} = \left\{\prod_{i=1}^{N} z_i\right\}^{1/N}, \quad (2.13)$$

and

$$\log \overline{g} = \frac{1}{N}\sum_{i=1}^{N} \log z_i, \quad (2.14)$$

in which the logarithm may be either natural (ln) or common ($\log_{10}$). If by transforming the data z_i, $i = 1, 2, \ldots, N$, we obtain $\log z$ with a normal distribution then the variable is said to be lognormally distributed. Its probability distribution is given by equation (2.9) in which z is replaced by $\ln z$, and σ and μ are the parameters on the logarithmic scale.

It is sometimes necessary to shift the origin for the transformation to achieve the desired result. If subtracting a quantity a from z gives a close approximation to normality, so that $z - a$ is lognormally distributed, then we have the probability density

$$f(z) = \frac{1}{\sigma(z-a)\sqrt{2\pi}} \exp\left[-\frac{1}{2\sigma^2}\{\ln(z-a) - \mu\}^2\right]. \tag{2.15}$$

We can write this as

$$f(z) = \frac{1}{\sigma(z-a)\sqrt{2\pi}} \exp\left[-\frac{1}{2\sigma^2}\left\{\ln\frac{z-a}{b}\right\}^2\right], \tag{2.16}$$

where $b = \exp(\mu)$. This is known as the three-parameter log-transformation; the parameters a, b and σ represent the position, size and shape, respectively, of the distribution. You can read more about this distribution in Aitchison and Brown (1957).

2.4.2 Square root transformation

Taking logarithms will often normalize, or at least make symmetric, distributions that are strongly positively skewed, i.e. have $g_1 > 1$. Less pronounced positive skewness can be removed by taking square roots:

$$r = \sqrt{z}. \tag{2.17}$$

2.4.3 Angular transformation

This is sometimes used for proportions in the range 0 to 1, or 0 to 100 if expressed as percentages. If p is the proportion then define

$$\phi = \sin^{-1}\sqrt{p}. \tag{2.18}$$

The desired transform is the angle whose sine is $\sqrt{p}$.

2.4.4 Logit transformation

If, as above, p is a proportion $(0 < p < 1)$, then its logit is

$$l = \ln\left\{\frac{p}{1-p}\right\}. \tag{2.19}$$

Note that the limits 0 and 1 are excluded; otherwise l would either go to $-\infty$ or $+\infty$. If you have proportions that include 0 or 1 then you must make some little adjustment to use the logit transformation.

In Chapter 10 we shall see a more elaborate transformation using Hermite polynomials.

2.5   EXPLORATORY DATA ANALYSIS AND DISPLAY

The physics of the environment might determine what transformation would be appropriate. More often than not, however, one must decide empirically by inspecting data. This is part of the preliminary exploration of the data from survey, which should always be done before more formal analysis. You should examine data by displaying them as histograms, box-plots and scatter diagrams, and compute summary statistics. You should suspect observations that are very different from their neighbours or from the general spread of values, and you should investigate abnormal values; they might be true outliers, or errors of measurement, or recording or transcription mistakes. You must then decide what to do about them.

If the data are not approximately normal then you can experiment with transformation to make them so, as outlined in Section 2.4. There are formal significance tests for normality, but these are generally not helpful, partly because they depend on the number of data and partly because they do not tell you in what way a distribution departs from normal. We illustrate this weakness below. You can try fitting theoretical distributions from the estimated parameters of the distribution to the histogram. If the histogram appears erratic then another way of examining the data for normality is to compute the cumulative distribution and plot it against the normal probability on normal probability paper. This paper has an ordinate scaled in such a way that a normal cumulative distribution appears as a straight line. Alternatively, you can compute the normal equivalent deviate for probability p; this is the value of z to the left of which on the graph the area under the standard normal curve is p. A strong deviation from the line indicates non-normality, and you can try drawing the cumulative distributions of transformed data to see which gives a reasonable fit to the line before deciding whether to transform and, if so, in what way.

To illustrate these effects we turn to the distribution of potassium at Broom's Barn Farm. The data are from an original study by Webster and McBratney (1987). The distribution is shown as a histogram of the measured values in Figure 2.1(a). To it is fitted the curve of the lognormal distribution with parameters as given in Table 2.1. It is positively skewed. The histogram of the logarithms is shown in Figure 2.1(b). It is approximately symmetric, the normal pdf fits well, and transforming to logarithms has approximately normalized the data. Figure 2.2 shows the corresponding box-plots, as 'box and whisker' plots in which the limits of the boxes enclose the interquartile ranges and the whiskers extend to the limits of the data, Figure 2.2(a)–(b). In Figure 2.2(c)–(d) the whiskers extend only to 'fences', and any points lying beyond them are plotted individually. The upper fence is the limit of the upper quartile plus 1.5 times the interquartile range or the maximum if that is smaller; the lower fence is defined analogously. Again, skew is seen to be removed by taking log-

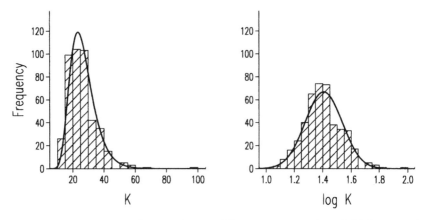

Figure 2.1 Histograms of (a) exchangeable potassium (K) and (b) $\log_{10}$ K, for the topsoil at Broom's Barn Farm. The curves are of the (lognormal) probability density.

arithms. Figure 2.3(a)-(b) shows the cumulative distributions plotted on the probability scale and as normal equivalent deviates, respectively. Figure 2.3(c)-(d) shows the same graphs for $\log_{10}$ K. These graphs are close to the normal line, and clearly transformation to logarithms yields a near-normal distribution in this instance.

Figures 2.4–2.6 show the effects of transformation to common logarithms for readily extractable copper of the topsoil in the Borders Region of Scotland (McBratney *et al.*, 1982). For these data, which are summarized in Table 2.2, taking logarithms normalizes the data very effectively.

The shortcomings of formal testing for a theoretical distribution can be seen in the χ^2 values given in Tables 2.1 and 2.2 for fitting the normal distribution. The values for the untransformed data are huge and clearly significant. Transforming potassium to logarithms still gives a χ^2 (43.6) exceeding the 5% value ($\chi^2_{p=0.05,\ f=18} = 28.87$), where p signifies the probability and f the degrees of freedom. Even for log Cu the computed χ^2 (28.1) is close to the 5% value. The reason, as mentioned above, lies largely in having so many data, so that the test is very sensitive.

2.5.1 Spatial aspects

For spatial data the spatial coordinates must also be checked. The positions of the sampling points can be plotted on a map, referred to in Chapter 1 as a 'posting' of the data. Do all the points lie within the region surveyed? If not, why? Sampling points for a soil survey falling in the sea are obviously wrong, but those on land just outside the region might be valid. Frequently the cause is a reversal of the coordinates, however.

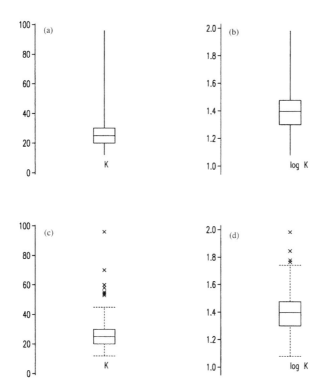

Figure 2.2 Box-plots of (a) exchangeable K and (b) $\log_{10} K$ showing the 'box' and 'whiskers', and (c) exchangeable K and (d) $\log_{10} K$ showing the fences at the quartiles plus and minus 1.5 times the interquartile range.

The data should also be examined for trend, which might be evident as a gross regional change in the values, which is also smooth and predictable. If you have sampled on a grid then arrange data in a two-way table, and compute the means and medians of both rows and columns, and plot them. The results will show if the data embody trend, at least in the directions of the axes of the coordinate system, by a progressive increase or decrease in the row or column means. Figure 2.7 shows the distribution of the sampling points for Broom's Barn Farm. The graphs of the row and column means are on the right-hand side and at the bottom, respectively. These graphs show small fluctuations about the row and column means, but no evidence of trend.

2.6   SAMPLING AND ESTIMATION

We have made the point above that we can rarely have complete information about the environment. Soil, for example, forms a continuous mantle

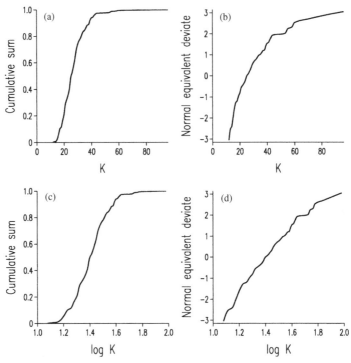

Figure 2.3   Cumulative distribution of (a) exchangeable K in the range 0 to 1 and (b) as normal equivalent deviates, on the original scale (mg l^{-1}), and (c) log$_{10}$ K in the range 0 to 1 and (d) as normal equivalent deviates.

Table 2.1   Summary statistics for exchangeable potassium (K, mg l^{-1}) at Broom's Barn Farm.

	K	log$_{10}$ K
Minimum	12.0	1.079 2
Maximum	96.0	1.982 3
Mean	26.31	1.398 5
Median	25.0	1.397 9
Standard deviation	9.039	0.134 2
Variance	81.706	0.018 00
Skewness	2.04	0.39
Kurtosis	9.51	0.57
Number of observations	434	434
χ^2 for normal fit (with 18 degrees of freedom)	174.4	43.6

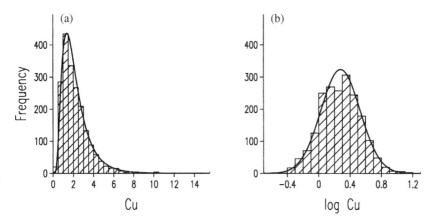

Figure 2.4 Histogram of (a) extractable copper (Cu) and (b) $\log_{10}$ Cu, in the top-soil of the Borders Region. The curves are of the (lognormal) probability density.

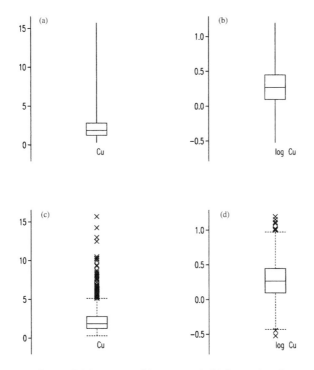

Figure 2.5 Box-plots of (a) extractable Cu and (b) $\log_{10}$ Cu showing the 'box' and 'whiskers', and (c) extractable Cu and (d) $\log_{10}$ Cu showing the fences at the quartiles plus and minus 1.5 times the interquartile range.

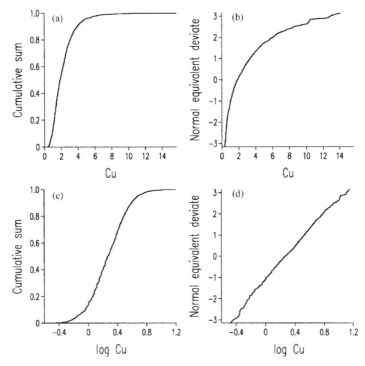

Figure 2.6 Cumulative distribution of (a) extractable Cu in the range 0 to 1 and (b) as normal equivalent deviates, on the original scale (mg kg^{-1}), and (c) log$_{10}$ Cu in the range 0 to 1 and (d) as normal equivalent deviates.

Table 2.2 Summary statistics for extractable copper (Cu, mg kg^{-1}) in the Borders Region.

	Cu	log$_{10}$ Cu
Minimum	0.3	−0.5214
Maximum	15.7	1.1959
Mean	2.221	0.2713
Median	1.85	0.2674
Standard deviation	1.461	0.2544
Variance	2.1346	0.064731
Skewness	2.52	0.06
Kurtosis	12.10	−0.05
Number of observations	1949	1949
χ^2 for normal fit (with 18 degrees of freedom)	977.6	28.1

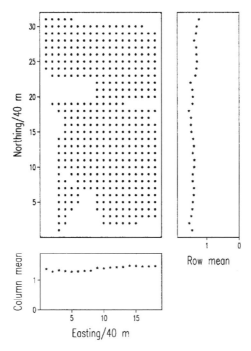

Figure 2.7 Posting of data for Broom's Barn Farm with the row and column means plotted on the right-hand side and at the bottom, respectively.

on the land except where it is broken by water or rock. Measurements, in contrast, are made on small cores or on bulked samples from small plots or fields. Similarly, rainfall is recorded in small gauges separated from one another by large distances. Data in this sense are fragmentary; they constitute a sample from whatever region is of interest, and from them we can try to describe the region in terms of mean values and variation.

The principal advances in sampling theory, sometimes known as classical theory, were made in the 1930s. The aim was to estimate means, and to a lesser extent higher-order moments, especially variances. It was not concerned to express spatial variation, which has become the province of geostatistics. Nevertheless, many of the ideas and formulae for geostatistics derive from the classical theory, and we therefore devote a short section to them. For fuller treatment you should consult one of the standard texts such as Cochran (1977) and Yates (1981).

2.6.1 Target population and units

The first step in sampling theory is to define a *target population*. This population comprises a set of *units*. In environmental survey a population is almost always circumscribed by the boundary of a physical region, and the units are all the places within it at which one might measure its properties. Measurements must be made on bodies of material with finite size, and so there is a finite number of non-overlapping units in the population. The units are usually so small in relation to the whole region that the population is effectively infinite. Millions of rain gauges 30 cm across could fit into a region of several hundred square kilometres without overlapping. The same is true of boreholes and soil profile pits. Even if the units were fields, there would be thousands of them. Nevertheless, they are all large enough to encompass variation, and in any one survey they should be of the same size. In fact, they should all have the same size, shape and orientation, known as the *support* of the sample.

The population is sampled by taking a subset of its units on a defined support. In classical theory this subset must be chosen with some element of randomization to ensure that the estimates from it are unbiased and to provide a probabilistic basis for inference. Perhaps paradoxically, the units must be selected according to a design to achieve this, and the technique is often called 'design-based estimation' in consequence.

2.6.2   Simple random sampling

This is the simplest form of design. Every unit in the sample is chosen without regard to any other, and all units have the same chance of selection.

Estimates from a simple random sample

If there are N units in the sample then its *mean*, $\bar{z}$, estimates the mean of the parent population, μ, by

$$\hat{\mu} = \bar{z} = \frac{1}{N} \sum_{i=1}^{N} z_i. \tag{2.20}$$

The *variance* of the population is the expected mean squared difference between μ and z, i.e. it is the mean of $(z - \mu)^2$, denoted by σ^2. It is estimated by

$$\hat{\sigma}^2 = s^2 = \frac{1}{N-1} \sum_{i=1}^{N} (z_i - \bar{z})^2. \tag{2.21}$$

The divisor is $N - 1$, not N, and this difference between the formula for the estimated variance of a population and the variance of a finite set, equation (2.3), arises because we do not know the true mean, but have only an estimate of it from the data. The *standard deviation* of the sample, s, computed using equation (2.21) estimates σ. In like manner we estimate the population covariance between two variables by replacing the divisor N in equation (2.10) by $N - 1$.

Estimation variance and standard error

All estimates are subject to error: sample information is never complete, and we want a measure of the uncertainty. This is usually expressed by the estimation variance of a mean:

$$s^2(\bar{z}) = \hat{\sigma}^2(\bar{z}) = s^2/N. \qquad (2.22)$$

It estimates the variance we should expect if we were to sample repeatedly and compute the average squared difference between the mean μ and the sample mean, $\bar{z}$:

$$\begin{aligned} \mathrm{E}[s^2(\bar{z})] &= \mathrm{E}[(\bar{z} - \mu)^2] \\ &= \sigma^2/N. \end{aligned} \qquad (2.23)$$

Its square root is the standard error, $s(\bar{z})$.

Naturally, $s^2(\bar{z})$ should be as small as possible. Evidently we can decrease $s^2(\bar{z})$, and improve our estimates, by increasing N, the size of the sample. Unless we can measure every unit in a population, however, we cannot eliminate the error. Further, simply increasing N confers less and less benefit for the effort involved, and beyond about 25 the gain in precision is disappointing.

2.6.3 Confidence limits

Having obtained an estimate and its variance we may wish to know within what interval it lies for any degree of confidence. If the variable has a normal distribution and the sample is reasonably large then the confidence limits for the mean are readily obtained as follows.

We consider a *standard normal deviate*, i.e. a normally distributed variable, y, with a mean of 0 and variance of 1, sometimes written $\mathcal{N}(0, 1)$. Then for any μ and σ

$$y = \frac{z - \mu}{\sigma}. \qquad (2.24)$$

Confidence limits on a mean are given by

$$\bar{z} - ys/\sqrt{N} \quad \text{and} \quad \bar{z} + ys/\sqrt{N}. \qquad (2.25)$$

These are the lower and upper limits on μ, given a sample mean $\bar{z}$ and standard deviation s that estimates σ^2 precisely, corresponding to some chosen probability or level of confidence. Values of standard normal deviates and their cumulative probabilities are published, and we list the values a few typical confidences at which people might wish to work and

Table 2.3 Typical confidences and their associated standard normal deviates, y.

Confidence (%)	68	75	80	90	95	99
y	1.0	1.15	1.28	1.64	1.96	2.58

the associated values of y in Table 2.3. The first entry is usually too liberal, and we include it only to show that approximately 68% of a normally distributed population lies within the range $-\sigma$ to $+\sigma$.

2.6.4 Student's t

With small samples s^2 is a poor estimate of σ^2, and in these circumstances one should replace y in expressions (2.25) by Student's t, which is defined by

$$t = \frac{\overline{z} - \mu}{s/\sqrt{N}}. \tag{2.26}$$

The true mean, μ, is unknown of course, but t has been worked out and tabulated for N up to 120. So one chooses the confidence level, and then finds from the published table the value of t corresponding to $N - 1$ *degrees of freedom*. The confidence limits of the mean are then

$$\overline{z} - ts/\sqrt{N} \quad \text{and} \quad \overline{z} + ts/\sqrt{N}. \tag{2.27}$$

As N increases so t approaches y, and for $N \geqslant 60$ the differences are trivially small. So we need use t only when $N < 60$.

2.6.5 The χ^2 distribution

Let $y_1, y_2, \ldots, y_m$ be m values drawn from a standard normal distribution. Their sum of squares is

$$\chi^2 = \sum_{i=1}^{m} y_i^2. \tag{2.28}$$

This quantity has the distribution

$$f(x) = \{2^{f/2}\Gamma(f/2)\}^{-1} x^{(f/2)-1} \exp(-x/2) \quad \text{for } x \geqslant 0, \tag{2.29}$$

where f is the number of degrees of freedom, equal to $N - 1$ in our case, and Γ is the gamma function defined for any $k > 0$ by

$$\Gamma(k) = \int_0^\infty x^{k-1} \exp(-x) \, dx.$$

Values of χ^2 have been worked out and tabulated, and can be found in any good book of statistical tables, such as that by Fisher and Yates (1963). They are also available in many statistical packages on computers.

The variance estimated from a sample is, from equation (2.21),

$$s^2 = \frac{1}{N-1} \sum_{i=1}^{N} (z_i - \mu)^2. \tag{2.30}$$

Dividing through by σ^2 gives

$$\frac{s^2}{\sigma^2} = \frac{1}{N-1} \sum_{i=1}^{N} \frac{(z_i - \mu)^2}{\sigma^2}, \tag{2.31}$$

and so

$$s^2/\sigma^2 = \chi^2/(N-1) \quad \text{and} \quad \chi^2 = (N-1)s^2/\sigma^2$$

with $N-1$ degrees of freedom, provided the original population was normally distributed.

Rearranging the last expression gives the following limits for a variance:

$$\frac{(N-1)s^2}{\chi^2_{p_1}} \leqslant \sigma^2 \leqslant \frac{(N-1)s^2}{\chi^2_{p_2}}, \tag{2.32}$$

where p_1 and p_2 are the probabilities and for which we can obtain values of χ^2 from the published tables.

2.6.6 Central limit theorem

In the foregoing discussion of confidence limits (Section 2.6.3) we have restricted the formulae to those for the normal distribution, the properties of which are so well established. It lends weight to our argument for transforming variables to normal if that is possible. However, even if a variable is not normally distributed it is often still possible to use the tabulated values and formulae when working with grouped data. As it happens, the distributions of sample means tend to be more nearly normal than those of the original populations. Further, the bigger is a sample the closer is the distribution of the sample mean to normality. This is the central limit theorem. It means that we can use a large body of theory when studying samples from the real world.

We might, of course, have to work with raw data that cannot readily be transformed to normal, and in these circumstances we should see whether the data follow some other known distribution. If they do then the same line of reasoning can be used to arrive at confidence limits for the parameters.

2.6.7 Increasing precision and efficiency

The confidence limits on means computed from simple random samples can be alarmingly wide, and the sizes of sample needed to obtain satisfactory precision can also be alarmingly large. One reason when sampling space with a simple random design is that it is *inefficient*. Its cover is uneven; there are usually parts of the region that are sparsely sampled while elsewhere there are clusters of sampling points. If a variable z is spatially *autocorrelated*, which is likely at some scale, then clustered points duplicate information. Large gaps between sampling points mean that information that could have been obtained is lacking. Consequently, more points are needed to achieve a given precision, as measured by $s^2(\bar{z})$, than if the points are spread more evenly. There are several better designs for areas, and we consider the two most common ones, *stratified random* and *systematic*.

Stratified sampling

In stratified designs the region of interest, R, is divided into small subdivisions (*strata*). These are typically small squares, but they may be other shapes, of equal area. At least two sampling points are chosen randomly within each stratum. For this scheme the largest possible gap is then less than four strata.

The variance within a stratum k is estimated from n_k data in it by

$$s_k^2 = \frac{1}{n_k - 1} \sum_{i=1}^{n_k} (z_{ik} - \bar{z}_k)^2, \tag{2.33}$$

in which z_{ik} are the measured values and $\bar{z}_k$ is their mean. If there are K strata then by averaging their variances we can obtain the estimation variance for the region:

$$s^2(\bar{z}, \text{stratified}) = \frac{1}{K^2} \sum_{k=1}^{K} \frac{s_k^2}{n_k}. \tag{2.34}$$

Its square root is the standard error.

The quantity $(1/k) \sum_{k=1}^{K} s_k^2$ is the pooled within-stratum variance, denoted by s_W^2. If there is any spatial dependence then it will be less than s^2, and so the variance and standard error of a stratified sample will be less than that of a simple random sample for the same effort, the same size of sample.

The ratio $s^2(\bar{z})/s^2(\bar{z}, \text{stratified})$ is the *relative precision* of stratification.

If we were happy with the precision achieved by simple random sampling then we could get the same precision by stratification with a smaller sample. Stratified sampling is more *efficient* by the factor

$$N_{\text{random}}/N_{\text{stratified}}.$$

Systematic sampling

Systematic sampling provides the most even cover. In one dimension the sampling points are placed at equal intervals along a line, a transect. In two dimensions the points may be placed at the intersections of an equilateral triangular grid for maximum precision or efficiency. With this configuration the maximum distance between any unsampled point and the nearest point on the sampling grid is the least. However, rectangular grids are more practical, and the loss of precision compared with triangular ones is usually so small that they are preferred.

The main disadvantage of systematic sampling is that classical theory provides no means of determining the variance or standard error from the sample because once one sampling point has been chosen (and the orientation in two dimensions) there is no randomization. An approximation may be obtained by dividing the region into strata and computing the pooled within-stratum variance as if sampling were random within the strata. The result will almost certainly be an overestimate, and conservative therefore. A closer approximation, and one that will almost certainly be close enough, can usually be obtained by Yates's method of *balanced differences* (Yates, 1981).

Estimates of error by balanced differences are computed as follows. Consider first regular sampling on a transect, i.e. in one dimension. The transect is viewed through a small window containing, say, m sampling points with values $z_1, z_2, \ldots, z_m$. We then compute for the window the differences:

$$d_m = \tfrac{1}{2}z_1 - z_2 + z_3 - z_4 + \cdots + \tfrac{1}{2}z_m. \tag{2.35}$$

A value of $m = 9$ is convenient. We then move the window along the transect in steps and compute d_m at each new position. If the transect is short then the positions should overlap; if not, a satisfactory procedure is to choose the first sampling point in a new position as the last one in the previous position. In this way every sampling point contributes, and with equation (2.35) all contribute equally. Then the variance for the transect mean is the sum

$$s^2 \text{ (balanced differences)} = \frac{1}{J(m - 2 + 0.5)} \sum_{j=1}^{J} d_{mj}^2, \tag{2.36}$$

where J is the number of steps or positions of the window, and the quantity $m - 2 + 0.5$ is the sum of the squares of the coefficients in equation (2.35).

For a two-dimensional grid the procedure is analogous. One chooses a square window. For illustration let it be of side 4. The coefficients can be assigned as follows:

−0.25	+0.5	−0.5	+0.25
+0.5	−1.0	+1.0	−0.5
−0.5	+1.0	−1.0	+0.5
+0.25	−0.5	+0.5	−0.25

The variance is calculated as in equation (2.36), now with the divisor $J \times 6.25$, the value 6.25 being the sum of the squares of the coefficients in the table above. Again, the positions of the window may overlap, but usually it is sufficient to arrange them so that only the sides are in common, and with this arrangement and the coefficients listed all points count and carry equal weight.

What these schemes do in both one and two dimensions, and in three if the scheme is extended, is to filter out long-range fluctuation, just as stratification does.

Where there is trend across the sampled region or periodicity, as, for example, in an orchard or as a result of land drainage, systematic sampling can give biased estimates of means. Such bias can be avoided by randomizing systematically within the grid. The result is *unaligned sampling* (see Webster and Oliver, 1990). It gives almost even cover. The disadvantage is the same as that of strict grid sampling in that the error cannot be estimated very accurately. The best procedure again is to stratify the region and compute the pooled within-stratum variance. Empirical studies have shown some big gains in precision and efficiency from both systematic and unaligned sampling (again, see Webster and Oliver, 1990, for an example).

2.6.8   Soil classification

Another way of stratifying a region to improve the precision of estimates is to divide it on the basis of certain attributes. This practice is widespread in land resource surveys, and it was the norm in soil survey. Soil surveyors stratify, i.e. classify, regions on the appearance of the soil in profile and on related features in the landscape.

Regional mean

If the classification is good then the within-class variance of a stratum, i.e. the pooled within-stratum variance, is smaller than the total variance. Classification should therefore improve the precision or efficiency in estimating the regional mean.

The classes of soil are rarely equal in area, and so the formula, equation (2.34), must be adjusted accordingly. We define a weight, w_k, for the kth stratum or class in proportion to the area it covers:

$$w_k = \frac{\text{area of stratum } k}{\text{total area}}.$$

The mean, μ, for the whole area is then estimated by the weighted average:

$$\bar{z} = \sum_{k=1}^{K} w_k \bar{z}_k, \tag{2.37}$$

where $\bar{z}_k$ is the estimated mean of the kth stratum. The estimation variance is

$$s^2(\bar{z}, \text{stratified}) = \left\{ \sum_{k=1}^{K} \frac{w_k^2 s_k^2}{n_k} \right\}. \tag{2.38}$$

The average within-class variance and other diagnostics of a classification can be estimated from data by *analysis of variance*, which is both elegant and powerful. It can also serve for prediction, and we therefore defer its treatment to the next chapter.

3

Prediction and interpolation

3.1 SPATIAL INTERPOLATION

As mentioned in Chapter 2, measurements of the environment, of soil, weather, rock, and water, are made on small bodies of material (supports) separated from one another by relatively large distances. They constitute a sample from a continuum that cannot be recorded everywhere. Yet the people who make the measurements or their clients would almost always like to know what the values are in the intervening space; they want to predict in a spatial sense from their more or less sparse data. For example, meteorologists want to predict rainfall from their rain gauges, hydrologists want to predict flow properties in rock from their measurements in boreholes, mining engineers want to estimate ore grades from diamond drill cores, and pedologists and agronomists want to estimate concentrations of elements in the soil from auger samples. Further, they usually want to map the spatial distributions of these variables. Their desires are almost as old the subjects themselves, and there have been many attempts to satisfy them quantitatively. They constitute the principal force driving geostatistics to meet practical needs; first in ore evaluation because of the huge costs of mining and metal extraction, but now in other branches of environmental science such as those we have listed.

Most attempts at spatial prediction have been mathematical, based on geometry and some appreciation of the physical nature of the phenomena. Most take account of only systematic or deterministic variation, but not of any error. In these respects, as we shall see, they fall short of what is needed practically. In some ways geostatistical prediction, kriging, is the logical conclusion of these attempts in that it builds on them and overcomes their weaknesses.

Nearly all the methods of prediction, including the simpler forms of kriging, can be seen as weighted averages of data. Thus we have the general prediction formula

$$z^*(\mathbf{x}_0) = \sum_{i=1}^{N} \lambda_i z(\mathbf{x}_i), \tag{3.1}$$

where $\mathbf{x}_0$ is a target point for which we want a value; the $z(\mathbf{x}_i)$, $i = 1, 2, \ldots, N$, at places $\mathbf{x}_i$ are the measured data; and λ_i are the weights assigned to them. For now we shall denote the prediction by $z^*(\mathbf{x}_0)$. First we examine how the weights are assigned for some of the common methods, and we leave kriging until Chapter 8 after we have dealt with its underlying theory.

3.1.1   Thiessen polygons (Voronoi polygons, Dirichlet tessellation)

This method is one of the earliest and simplest. The region sampled, R, is divided by perpendicular bisectors between the N sampling points into polygons or tiles, V_i, $i = 1, 2, \ldots, N$, such that in each polygon all points are nearer to its enclosed sampling point $\mathbf{x}_i$ than to any other sampling point. The prediction at each point in V_i is the measured value at $\mathbf{x}_i$, i.e. $z^*(\mathbf{x}_0) = z(\mathbf{x}_i)$. The weights are

$$\lambda_i = \begin{cases} 1 & \text{if } \mathbf{x}_i \in V_i, \\ 0 & \text{otherwise.} \end{cases} \tag{3.2}$$

The shortcomings of the method are evident; each prediction is based on just one measurement, there is no estimate of the error, and information from neighbouring points is ignored. When used for mapping the result is crude; the interpolated surface consists of a series of steps.

3.1.2   Triangulation

Another early group of interpolators comprises those deriving from triangulation. The sampling points are linked to their neighbours by straight lines to create triangles that do not contain any of the points. The measured values are envisaged as standing above the basal plane at a height proportional to those values so that the whole set of data forms a polyhedron consisting of more or less tilted triangular plates. The aim is to determine the height of the plate at $\mathbf{x}_0$ from the apices of the triangle by linear interpolation.

This can be represented as a weighted average with weights determined as follows. We denote the coordinates of the three apices by $\{x_{11}, x_{12}\}$, $\{x_{21}, x_{22}\}$ and $\{x_{31}, x_{32}\}$ and those of the target point by $\{x_{01}, x_{02}\}$. Then the weights are given by

$$\lambda_1 = \frac{(x_{01} - x_{31})(x_{22} - x_{32}) - (x_{02} - x_{32})(x_{21} - x_{31})}{(x_{11} - x_{31})(x_{22} - x_{32}) - (x_{12} - x_{32})(x_{21} - x_{31})}, \tag{3.3}$$

with analogous expressions for λ_2 and λ_3. All other weights are 0.

The technique is simple and local. The disadvantages are that, although it is somewhat better than the Thiessen method, each prediction still depends on only three data; it makes no use of data further away, and there is again no measure of error. Unlike the Thiessen method, the resulting surface is continuous, but it has abrupt changes in gradient at the margins of the triangles. If the principal aim is to predict rather than to make a map with smooth isolines then the discontinuities in the derivative are immaterial. Another difficulty is that there is no obvious triangulation that is better than any other; even for a rectangular grid there are two options.

3.1.3 Natural neighbour interpolation

Sibson (1981) combined the best features of the two methods above in what he called 'natural neighbour interpolation'. The first step is a triangulation of the data by Delauney's method in which the apices of the triangles are those sampling points in adjacent Dirichlet tiles. This triangulation is unique except where the data are on a regular rectangular grid. To determine the value at any other point, x_0, that point is inserted into the tessellation, and its neighbours, the set T (the points within its bounding Dirichlet tiles), are used for the interpolation. Sibson called these points 'natural neighbours'.

For each neighbour the area, A, of the portion of its original Dirichlet tile that became incorporated in the tile of the new point is calculated. These areas, when scaled to sum to 1, become the weights. We can represent this by the general formula:

$$\lambda_i = \frac{A_i}{\sum_{k=1}^{N} A_k} \quad \text{for all } i = 1, 2, \ldots, N. \quad (3.4)$$

This means that if a point x_i is a natural neighbour, i.e. $x_i \in T$, then A_i has a value and the point carries a positive weight. If x_i is not a natural neighbour then it has no area in common with the target and its weight, λ_i, is zero.

This interpolator is continuous and smooth except at the data points where its derivative is discontinuous. Sibson called it the *natural neighbour C^0 interpolant*.

He did not like abrupt change in the surface at the data points, and so he elaborated the method by calculating the gradients of the statistical surface at these from their natural neighbours. These gradients were then combined with the weighted measurements to provide the height at the new point. The result is a smooth, once differentiable surface. Like the simple polyhedral interpolator, it returns the actual values at the measured points, i.e. it is an exact interpolator. Sibson showed that it reproduces continuous mathematical functions faithfully. However, both we and Laslett *et al.* (1987) have found that it produces unacceptable results where data are noisy. At local maxima and minima in such data it generates 'Prussian helmets', which Sibson wished to avoid.

3.1.4 Inverse functions of distance

Somewhat more elaborate than triangulation, and much more popular, are the methods based on inverse functions of distance in which the weights are defined by

$$\lambda_i = 1 / |\mathbf{x}_i - \mathbf{x}_0|^\beta \quad \text{with } \beta > 0, \tag{3.5}$$

and again scaled so that they sum to 1. The result is that data points near to the target point carry larger weight than those further away. The most popular choice of β is 2 so that the data are inversely weighted as the square of distance. As with triangulation, if $\mathbf{x}_0$ coincides with any $\mathbf{x}_i$ then λ_i becomes infinite, the other weights are immaterial, and $z(\mathbf{x}_0)$ takes the value $z(\mathbf{x}_i)$. Interpolation is exact. An attractive feature of weighting by inverse squared distance is that the relative weights diminish rapidly as the distance increases, and so the interpolation is sensibly local. Further, because the weights never become zero there are no discontinuities. Its disadvantages are that the choice of the weighting function is arbitrary, and there is no measure of error. Further, it takes no account of the configuration of the sampling. So where data are clustered two or more may be at approximately the same distance and direction from $\mathbf{x}_0$, and each point will carry the same weight as an isolated point a similar distance away but in a different direction. This is clearly undesirable, and some implementations for mapping have elaborated the scheme to overcome this—see, for example, Shepard's (1968) solution in the once popular SYMAP program. The interpolated surface will have a gradient of zero at the data points, and maxima and minima can occur only there.

3.1.5 Trend surfaces

A method that became popular among earth scientists, especially petroleum geologists, when they first had access to computers was trend surface analysis. This is a form of multiple regression in which the predictors are the spatial coordinates. For example,

$$z(x_1, x_2) = f(x_1, x_2) + \varepsilon, \tag{3.6}$$

where $z(x_1, x_2)$ is the predicted value at $\{x_1, x_2\}$ and f denotes a function of the spatial coordinates there. The model contains an error term, ε, and in regression this is assumed to be independently and identically distributed with mean $= 0$ and variance σ_ε^2. Plausible functions, usually simple polynomials such as planes, quadratics or cubics, are fitted by least squares to the spatial coordinates, and the resulting regression equation is used for the prediction. Thus for a plane the regression equation would be

$$z = b_0 + b_1 x_1 + b_2 x_2, \tag{3.7}$$

and for a quadratic surface

$$z = b_0 + b_1 x_1 + b_2 x_2 + b_3 x_1^2 + b_4 x_2^2 + b_5 x_1 x_2. \tag{3.8}$$

The predictor can be expressed as a weighted average of the data used to obtain the trend surface as follows. We represent the spatial coordinates and their powers by a matrix $\mathbf{X}$ with N rows for the N sampling points and as many columns as coefficients b to be estimated. For a first-order surface we can write the spatial coordinates as the matrix

$$\mathbf{X} = \begin{bmatrix} 1 & x_{11} & x_{12} \\ 1 & x_{12} & x_{22} \\ \vdots & \vdots & \vdots \\ 1 & x_{N2} & x_{N2} \end{bmatrix},$$

in which the first column is a dummy variate of 1s, and the recorded values of z at those places as the vector

$$\mathbf{z} = \begin{bmatrix} z(\mathbf{x}_1) \\ z(\mathbf{x}_2) \\ \vdots \\ z(\mathbf{x}_N) \end{bmatrix}.$$

The coefficients b are obtained from the matrix multiplication

$$\mathbf{b} = (\mathbf{X}^\mathsf{T}\mathbf{X})^{-1}\mathbf{X}^\mathsf{T}\mathbf{z}, \tag{3.9}$$

and the predictions are then given by

$$z_0^* = \mathbf{x}_0 \mathbf{b}, \tag{3.10}$$

in which $\mathbf{x}_0$ is the row vector $[1 \; x_{01} \; x_{02}]$. Thus the weights are given by equation (3.9). For a more complex surface the matrix $\mathbf{X}$ is simply extended by adding columns for the additional powers of x_1 and x_2.

Initially trend surfaces seemed attractive, but enthusiasm soon turned to disappointment. In most instances spatial variation is so complex that a polynomial of very high order is needed to describe it, and the resulting matrix equations are usually unstable. The residuals from the trend are autocorrelated, and so one of the assumptions of regression is violated. As a consequence the errors calculated by the usual formula, such as equation (6.49) in Webster and Oliver (1990), are incorrect. The random component is often large and masks the deterministic trend, and fitting in one part of a region affects the predictions everywhere. Thus, in a region containing both mountains and plain the prediction of topographic height on the plain will be determined by the much larger fluctuations in the mountains. Trend surfaces are not sufficiently local, and they do not return the values at data points.

Nevertheless, simple regression surfaces can represent long-range trend in some instances. The technique has its merits therefore in revealing long-range structures and filtering them to leave variation of shorter range that can be analysed by other techniques—see Moffat *et al.* (1986) for an example.

3.1.6 Splines

A spline function also consists of polynomials, but each polynomial of degree p is local rather global. The polynomials describe pieces of a line or surface, and they are fitted together so that they join smoothly, and their $p - 1$ derivatives are continuous. The places at which the pieces join are known as 'knots', and the choice of knots confers an arbitrariness on the technique. Splines can be constrained to pass through the data. Alternatively by choosing knots away from the data points they can be fitted by least squares or some other method to produce smoothing splines. Typically the splines are of degree 3; these are cubic splines.

3.2 SPATIAL CLASSIFICATION AND PREDICTING FROM SOIL MAPS

We conclude this chapter with a look at prediction using spatial classification. Surveyors in most branches of environmental science divide the regions that they study into classes by boundaries, and they characterize each class so derived from sample data. The maps, choropleth maps to give them their technical name, are commonplace. The soil map showing a patchwork of colour is a familiar one, as are similar maps in geology and ecology. The intention, usually implied rather than expressed, is that the characteristic information for any one class, any one colour on the map, may be used to predict conditions elsewhere in the same class. Rarely, however, is this put in statistical terms, yet it is the only sound application of classical statistics to spatial prediction. Indeed, it combines classical survey in such fields as geology and soil science with classical statistics. Several scientists have analysed soil maps in this framework—for example, Kantey and Williams (1962), Morse and Thornburn (1961), and Webster and Beckett (1970)—but as far as we know, the subject is not covered in any textbook. We therefore describe it in some detail, drawing on a recent analysis by Leenhardt *et al.* (1994).

3.2.1 Theory

We start with a region of interest, R, that has been classified into K classes, separated by boundaries. Every point in R, i.e. every unit of the population, belongs to one and only one class.

For any class k in R we can express the value of any point $\mathbf{x}_i$, i.e. any unit i, selected at random as

$$Z_{ik} = \mu + \alpha_k + \varepsilon_{ik}, \tag{3.11}$$

where Z_{ik} is the value of z at $\mathbf{x}_i$ in class k, μ is the general mean of z, α_k is the difference between μ and the mean of the class k, and ε_{ik} is a random component with mean zero and variance σ_k^2, the variance within class k.

The mean of class k, $\mu_k = \mu + \alpha_k$, is estimated from n_k observations by

$$\hat{\mu}_k = \frac{1}{n_k} \sum_{i=1}^{n_k} z_i. \tag{3.12}$$

with variance

$$\sigma^2(\hat{\mu}_k) = \sigma_k^2 / n_k. \tag{3.13}$$

In the absence of other information μ_k is also the best predictor of z at any $\mathbf{x}_i$, $i \in k$, and in keeping with our general formula for linear predictors, equation (3.1), we can represent it as

$$z^*(\mathbf{x}_0) = \hat{\mu}_k = \sum_{i=1}^{N} \lambda_i z(\mathbf{x}_i), \tag{3.14}$$

in which

$$\lambda_i = \begin{cases} 1/n_k & \text{for } \mathbf{x}_i \in k, \\ 0 & \text{otherwise.} \end{cases} \tag{3.15}$$

Its prediction variance is the expected mean squared difference (MSE) between the true value and the predicted one:

$$\text{MSE}_k = E_i[\{Z_{ik} - \mu_k\}^2] = \sigma_k^2. \tag{3.16}$$

In practice we never know μ_k; we only ever have an estimate, $\hat{\mu}_k$. So its variance, var$[\hat{\mu}_k]$, is an additional source of error in our prediction. Further, there is the possibility that our estimate of μ_k is biased. So a term representing the bias should be added, and the full squared prediction error becomes

$$\text{MSE}_k = \sigma_k^2 + \text{var}[\hat{\mu}_k] + \text{bias}^2[\hat{\mu}_k]. \tag{3.17}$$

If we sample randomly and estimate μ_k by the arithmetic average of n_k, the observed values in k, then there is no bias, and var$[\hat{\mu}_k]$ equals σ_k^2 / n_k. Equation (3.17) then becomes

$$\text{MSE}_k = \sigma_k^2 + \sigma_k^2 / n_k = \sigma_k^2 (1 + 1/n_k). \tag{3.18}$$

An immediate practical problem is to estimate σ_k^2. Confidence limits on variances are typically wide (Chapter 2) for small samples, and for a map with many classes surveyors rarely have the resources to record at more than a few sites within each class. Consequently equation (3.18) is likely to lead to crude estimates of the MSE. To solve this problem we therefore make a further assumption, namely that the variance within classes is the same for all. In conventional soil mapping, for example, surveyors try to maintain the same categoric level for all the classes in any one survey, say, all soil series or all soil families. The intention, expressed or implied, is that classes are equally variable. In these circumstances σ_k^2 in the above equations may be replaced by σ_W^2, the average or pooled within-class variance. Our task now is to estimate it, and this is best done by an analysis of variance (see Chapter 2).

The total variance of Z in the region, designated by σ_T^2, can be written as

$$\sigma_T^2 = \sigma_W^2 + \sigma_B^2, \tag{3.19}$$

where σ_B^2 is the between-class variance. These quantities immediately lead to expressions of the efficacy of a classification at partitioning the variance of Z by, for example, the intraclass correlation:

$$\rho_i = \frac{\sigma_B^2}{\sigma_B^2 + \sigma_W^2} = 1 - (\sigma_W^2 / \sigma_T^2). \tag{3.20}$$

Evidently, the larger is σ_B^2 and the smaller is σ_W^2, and hence the larger is ρ_i, the better we should regard the classification.

Prediction using a random sample

If we sample a region by selecting points at random and with numbers proportional to the areas covered by the classes, then σ_W^2 and σ_B^2 are estimated by s_W^2 and s_B^2, respectively, without bias in a one-way analysis of variance. This leads equally to unbiased estimate of ρ_i by r_i (Webster and Beckett, 1968). Alternatively, we may take s_T^2 as an estimate of σ_T^2 and compute $R_i^2 = 1 - (s_W^2 / s_T^2)$, which is the proportion of variance in the data explained by the classification and analogous to R^2 in regression analysis.

We can now insert the pooled within-class variance, σ_W^2, into equation (3.18) in place of σ_k^2 to obtain

$$\text{MSE}_k = \sigma_W^2 (1 + 1/n_k). \tag{3.21}$$

Further, we can compute an average prediction error, $\overline{\text{MSE}}$, for the region by

$$\overline{\text{MSE}} = \sum_{k=1}^{K} A_k \sigma_W^2 (1 + 1/n_k), \tag{3.22}$$

where A_k is the proportion of the area of class k in R.

Prediction from a purposively chosen sample

Consider now predicting Z from a purposively chosen representative profile, p, in class j with value z_{pj}. The latter replaces $\hat{\mu}_j$ as an estimate of Z_{ij}. It is fixed, however, so $\text{var}[z_{pj}] = 0$, and the difference $d_j = z_{pj} - \mu_j$ is the bias of equation (3.17). The prediction variance is

$$\text{MSE}_{pj} = \sigma_j^2 + d_j^2. \tag{3.23}$$

Under the assumption of a common within-class variance, we obtain the expected mean squared error of prediction from class representatives for the whole region by

$$\overline{\text{MSE}}_p = E_j[\text{MSE}_{pj}] = \sigma_W^2 + \sum_{j=1}^{J} A_j d_j^2. \tag{3.24}$$

The minimum value of $\overline{\text{MSE}}_p$ is σ_W^2, which is reached when the $z_{pj} = \mu_j$ for all j.

3.2.2 Summary

Whether we predict Z using the means of random samples or from purposively chosen representatives, σ_W^2 sets a lower limit to the mean squared error of prediction. In the former case we can approach this minimum by increasing the size of sample; in the latter by selecting the representatives to match the mean values as closely as possible. If we want to improve prediction further using the conventional approach, we must diminish the within-class variance by refining the classification. This might be done by increasing the scale so that boundaries can be delineated more accurately and intricately, or by subdividing the soil more finely, i.e. by increasing the number of classes. In practice the second is likely to demand the first: there is no point in creating classes that cannot be displayed at the chosen scale. Alternatively, we might devise a special classification for each property we wish to predict.

The effectiveness of the conventional procedure for soil survey depends both on the quality of the classification and its mapping and on the ability of the surveyor to select representative soil profiles in the field, where the values of the properties of interest approximate the class means. In particular, $\overline{\text{MSE}}_p$ should be less than $2\sigma_W^2$, otherwise the selection is worthless, and $2\sigma_W^2$ should be less than $\sigma_T^2 + \sigma_T^2/N$, where N is the total size of sample, otherwise classification confers no benefit.

For the whole procedure to be successful we want

$$\sigma_W^2 < \overline{\text{MSE}}_p < 2\sigma_W^2 < \sigma_T^2 + \sigma_T^2/N. \tag{3.25}$$

To complete the picture we have to estimate $\overline{\text{MSE}}_p$. Let us assume that we have for each class j, $j = 1, 2, \ldots, J$, one representative with value z_{pj} and $V(j)$ validation points chosen probabilistically with values Z_{vj}, $v = 1, 2, \ldots, V(j)$, and that there are N_v validation points in all. Then

$$\hat{d}_p = \frac{1}{N_v} \sum_{j=1}^{J} \sum_{v=1}^{V(j)} (Z_{vj} - z_{pj}) \tag{3.26}$$

and

$$\widehat{\text{MSE}}_p = \frac{1}{N_v} \sum_{j=1}^{J} \sum_{v=1}^{V(j)} (Z_{vj} - z_{pj})^2. \tag{3.27}$$

4

Characterizing spatial processes: the covariance and variogram

4.1 INTRODUCTION

The previous chapter describes several common methods of spatial interpolation. Some of them are crude, so that maps made using them display the spatial variation poorly. The interpolators also fail to provide any estimates of the error, which are desirable for prediction. The conventional approach to spatial prediction in soil science combines classical estimation with spatial classification and thereby overcomes some of these weaknesses. It is the only method described in Chapter 3 that gives sound estimates of error. However, it requires replicated sampling for each class to provide individual estimates for that class and some degree of randomization of the sample. The sampling effort can be large, but even with such effort the predicted values at all points within a given class are simply the mean of that class for the property of interest. The precision of prediction is limited by the goodness of the classification; variation within classes is ignored, and local variation is not resolved.

Mathematical functions of the spatial coordinates seemed at one time to have promise. They could be defined fully, and they could therefore be used repeatably. Most were also intuitively reasonable. Some, such as the inverse functions of distance and triangulation, however, were also quite arbitrary, taking no account of more general knowledge of the variation in the region. Trend surface analysis, the only function described in Chapter 3 that does recognize the generality, has other defects.

The methods are deterministic, and to that extent they accord with our understanding that the variation in the environment has physical causes, i.e. is physically determined. However, the environment and its component attributes, such as the soil, result from many physical and biological processes that interact, some in highly non-linear and chaotic ways.

The outcome is so complex that the variation appears to be random. This complexity, together with our current, far from complete, understanding of the processes, means that mathematical functions are not adequate to describe any but the simplest components.

A fully deterministic solution to our problems seems out of reach at present. To make progress we must look at spatial variation differently. Recapitulating, we have two needs: to describe quantitatively how soil varies spatially, and to predict its values at places where we have not sampled. In addition, we want estimates of the errors on these predictions so that we can judge what confidence to place in them; estimates of errors are lacking in the classical methods of interpolation. We need a model for prediction, and since there is no deterministic one the solution seems to lie in a probabilistic or stochastic approach.

4.2   A STOCHASTIC APPROACH TO SPATIAL VARIATION: THE THEORY OF REGIONALIZED VARIABLES

4.2.1   Random variables

The fact that spatial variation appears to be random suggests a way forward. Consider throwing a die; on any one throw we obtain a number, for instance, a 6. This is the outcome of throwing the die once, of drawing one value from a distribution that consists of the set $\{1, 2, 3, 4, 5, 6\}$ with equal probability. One can argue that the result is physically determined in that it depends on the position of the die in the cup and of the cup itself at the start, the forces imparted to it by the thrower, and the nature of the surface on which it lands (Matheron, 1989). Nevertheless, these are so imperfectly known and so far beyond our control that we regard the process as random and as unbiased. Similarly, since the factors that determine the values of environmental variables are numerous, largely unknown in detail, and interact with a complexity that we cannot disentangle, we can regard their outcomes as random.

If we adopt a stochastic view then at each point in space there is not just one value for a property but a whole set of values. We regard the observed value there as one drawn at random according to some law, from some probability distribution. This means that at each point in space there is variation, a concept that has no place in classical estimation. Thus, at a point $\mathbf{x}$ a property, $Z(\mathbf{x})$, is treated as a random variable with a mean, μ, a variance, σ^2, and higher-order moments, and a cumulative distribution function (cdf). It has a full probability distribution, and it is from this that the actual value is drawn. If we know approximately what that distribution might be we can estimate values at unrecorded places from data in the neighbourhood and put errors on our estimates.

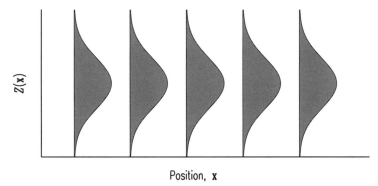

Figure 4.1   The normal distributions of the random variables at five sites.

Most environmental variables, such as the soil's pH and potassium concentration, are continuous. For these a value $z(\mathbf{x})$ can be thought of as one of an infinite number of possible values, with a cdf that is the probability that Z takes any value less than or equal to a particular value z_c:

$$F\{Z(\mathbf{x};z)\} = \text{Prob}[Z(\mathbf{x}) \leqslant z_c] \quad \text{for all } z. \tag{4.1}$$

The probability $F\{Z(\mathbf{x};z)\}$ takes values between 0 and 1. Its derivative is the probability density function, the pdf:

$$f\{Z(\mathbf{x})\} = \frac{\mathrm{d}F\{Z(\mathbf{x};z)\}}{\mathrm{d}z}, \tag{4.2}$$

which we described in Chapter 2. The distribution may be bounded, as in the case of a proportion or percentage, but the most useful assumption is that it is not, so that $-\infty \leqslant Z(\mathbf{x}) \leqslant +\infty$.

4.2.2   Random functions

The description above for an individual point $\mathbf{x}$ applies to the infinitely many in the space; at each point $\mathbf{x}_i$, $i = 1, 2, \ldots$, $Z(\mathbf{x}_i)$ has its own distribution and cdf. The range of possible values constitutes an *ensemble*, and one member of the ensemble is the realization. The idea is illustrated in Figure 4.1 in which the curves are imagined to protrude vertically out of the plane of the page. The set of random variables, $Z(\mathbf{x}_1), Z(\mathbf{x}_2), \ldots$, constitute a *random function*, a *random process*, or a *stochastic process*. The set of actual values of Z that comprise the realization of the random function are known as a *regionalized variable*. Just as in Chapter 2 we regarded a region as made up of a population of units, so we can think of a random function $Z(\mathbf{x})$ as a superpopulation, with an infinite number of units in space and an infinite number of values of Z at each point in the space. It is doubly infinite.

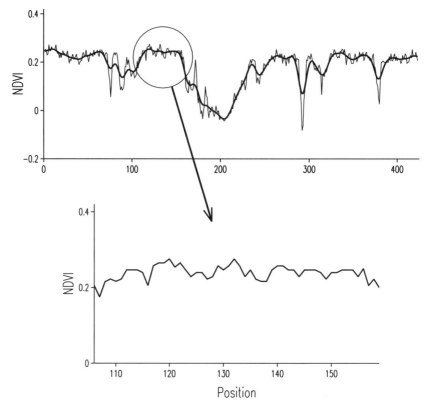

Figure 4.2   Transect across a SPOT image for normalized difference vegetation index. The fine line in the upper graph joins the data, and the bold line is a smoothing spline fitted through them. The lower graph is an enlarged version of the section in the circle.

4.3   SPATIAL COVARIANCE

To define the variation we need to describe the ensemble simply. For the possible outcomes of throwing the die it is easy because they are independent. The values of regionalized variables, on the other hand, tend to be related. In general, values at two places near to one another are similar, whereas those at more widely separated places are less so. This can be seen in Figure 4.2, which represents pixel values for the normalized difference vegetation index (NDVI), where

$$\text{NDVI} = (\text{infrared} - \text{red}) / (\text{infrared} + \text{red}),$$

along one row of a SPOT (Système Probatoire de l'Observation de la Terre) image (from Oliver *et al.*, 2000). The fine line joining the pixel values in the upper graph illustrates the locally erratic nature of the variation. Wherever we look we see some fluctuation, but in most short sections of the transect the values are similar. Over longer distances, however, the values vary more substantially, with some sections having small values on average and others where they are large. This becomes clear when the locally erratic variation has been filtered by a smoothing spline, the bold line in the graph. Where the property is continuous, as in this example and as is the case for most properties of the environment, its values must be related at some scale. This is illustrated further in the lower graph where a small section of the transect is magnified and the pixel values are plotted in more detail. What appeared to be entirely erratic in the upper graph can be seen at the larger resolution as structured in the sense that neighbouring values are similar to one another on average. We want to describe these relations, and we do so using the concept of covariance.

We are likely to be familiar with using the covariance to determine the relation between two variables for paired observations. For n pairs of observations, $z_{i,1}$, $z_{i,2}$, $i = 1, 2, \ldots, n$, of two variables, z_1 and z_2, the covariance is given by

$$\hat{C}(z_1, z_2) = \frac{1}{n} \sum_{i=1}^{n} \{z_{i,1} - \overline{z}_1\}\{z_{i,2} - \overline{z}_2\}, \tag{4.3}$$

where $\overline{z}_1$ and $\overline{z}_2$ are the means of z_1 and z_2, respectively. If the units $i = 1, 2, \ldots, n$, on which the observations were made were drawn at random then $\hat{C}(z_1, z_2)$ estimates the population covariance without bias.

We can extend this definition for relating two random variables. The concept and its mathematical expression were developed originally for analysing time series during the 1920s and 1930s, and they have been much used for processing signals and for forecasting. They are now described in many textbooks, of which we can recommend Jenkins and Watts (1968) and Priestley (1981). They have their analogies in space, and Yaglom (1987) presents them in this context as underpinning spatial prediction.

In our new spatial setting z_1 and z_2 become $Z(\mathbf{x}_1)$ and $Z(\mathbf{x}_2)$, i.e. they are the sets of values of the same property, Z, at the two places $\mathbf{x}_1$ and $\mathbf{x}_2$, and we have switched the notation to capital Z to signify that they are random variables. Their covariance is

$$C(\mathbf{x}_1, \mathbf{x}_2) = E[\{Z(\mathbf{x}_1) - \mu(\mathbf{x}_1)\}\{Z(\mathbf{x}_2) - \mu(\mathbf{x}_2)\}], \tag{4.4}$$

where $\mu(\mathbf{x}_1)$ and $\mu(\mathbf{x}_2)$ are the means of Z at $\mathbf{x}_1$ and $\mathbf{x}_2$. The equation is analogous to equation (4.3). Unfortunately, however, its solution is unavailable because we have only the one realization of Z at each point: we cannot know the means. Thus we seem to have reached an impasse, and we can progress only by making further assumptions of *stationarity* which allow us to treat the values at different places as though they are different realizations of the property.

4.3.1 Stationarity

By stationarity we mean that the distribution of the random process has certain attributes that are the same everywhere. Starting with the first moment, we assume that the mean, $\mu = E[Z(\mathbf{x})]$, about which individual realizations fluctuate, is constant for all $\mathbf{x}$. This enables us to replace $\mu(\mathbf{x}_1)$ and $\mu(\mathbf{x}_2)$ by the single value μ, which we can estimate by repetitive sampling.

We next consider the second moments. Equation (4.4) as written is restricted to the two particular points $\mathbf{x}_1$ and $\mathbf{x}_2$, which is not very useful. We want to generalize it so that it describes the process, and to do so we must make further assumptions. The first concerns what happens when $\mathbf{x}_1$ and $\mathbf{x}_2$ coincide. Equation (4.4) then defines the variance, $\sigma^2 = E[\{Z(\mathbf{x}) - \mu\}^2]$, sometimes called the *a priori* variance of the process. We assume this to be finite and, like the mean, to be the same everywhere. Second, when $\mathbf{x}_1$ and $\mathbf{x}_2$ do not coincide their covariance depends on their separation and not on their absolute positions: this applies to any pair of points $\mathbf{x}_i, \mathbf{x}_j$ separated by the vector $\mathbf{h} = \mathbf{x}_i - \mathbf{x}_j$, so that we have

$$C(\mathbf{x}_i, \mathbf{x}_j) = E[\{Z(\mathbf{x}_i) - \mu\}\{Z(\mathbf{x}_j) - \mu\}], \qquad (4.5)$$

which is constant for any given $\mathbf{h}$. This constancy of the mean, variance and covariances that depend only on separation and not on absolute positions, i.e. constancy of the first and second moments of the ensemble or process, constitutes *second-order stationarity* or *weak stationarity*. Note that the moments are of the imaginary random process of which we have the one realization and that we can never know their values exactly. We can estimate them, and formulae for doing that are given in the next chapter, equations (5.7) and (5.9).

Just as each random function has its cdf, so each pair of random functions $Z(\mathbf{x}_i)$ and $Z(\mathbf{x}_j)$ will have a joint cdf:

$$F\{Z(\mathbf{x}_i, \mathbf{x}_j; z)\} = \text{Prob}[Z(\mathbf{x}_i) \leqslant z, Z(\mathbf{x}_j) \leqslant z] \qquad \text{for all } z, \qquad (4.6)$$

and a corresponding pdf, the derivative of equation (4.6). Chapter 2 gives the formula for the pdf of a bivariate normal distribution. As an example, if we have a set of points regularly spaced along a line at positions $x_1, x_2, \ldots, x_N$, then we expect the joint cdf $F\{Z(x_1, x_2; z)\}$ to be the same as $F\{Z(x_2, x_3; z)\}$, as $\ldots$, and as $F\{Z(x_{N-1}, x_N; z)\}$. Further, it enables us to obtain a picture of the joint distribution of pairs of points one interval apart by sampling at these positions and plotting their values on a scatter diagram as a representation of the pdf. This is described in detail in the next chapter and illustrated in Figure 5.2. There will be $N - 1$ pairs one interval apart, $N - 2$ pairs two intervals apart, and $N - h$ pairs h intervals apart. In two and three dimensions the separation is a vector with both distance and direction, which we denote by $\mathbf{h}$, and is known as the *lag*.

The joint cdf will have higher-order moments. If these also depend on the separation only, then the process is said to be strictly or fully stationary. It is not always wise to assume such strong stationarity, but in practice it might not matter. If the distribution is normal (Gaussian) then the moments of order 3 and more are known constants, and we need not concern ourselves with them. This is another motivation for transforming non-normal data to normality if possible. Therefore, we can usually limit ourselves to nothing more demanding than second-order stationarity and concentrate on the covariance.

4.3.2 Ergodicity

Ergodicity is closely related to stationarity. A process is said to be ergodic when the moments of the single observable realization in space approach those of the ensemble as the regional bounds expand towards infinity. It is of mainly theoretical interest rather than of practical value because the regions we study are finite, and we never know the ensemble averages. Nevertheless, we sometimes have to distinguish, especially when choosing estimators.

4.4 THE COVARIANCE FUNCTION

We can rewrite equation (4.5) as

$$\text{cov}[Z(\mathbf{x}), Z(\mathbf{x} + \mathbf{h})] = \text{E}[\{Z(\mathbf{x}) - \mu\}\{Z(\mathbf{x} + \mathbf{h}) - \mu\}]$$
$$= \text{E}[\{Z(\mathbf{x})\}\{Z(\mathbf{x} + \mathbf{h})\} - \mu^2]$$
$$= C(\mathbf{h}). \tag{4.7}$$

In words, the covariance is a function of the lag, $\mathbf{h}$, and the lag only. It is the *autocovariance function—auto* because it represents the covariance of Z with itself. Unless there is any ambiguity, we shall refer to it simply as the covariance function. It describes the dependence between values of $Z(\mathbf{x})$ with changing lag. If $Z(\mathbf{x})$ has a multivariate normal distribution for all positions then the mean and the covariance function completely characterize the process because all of the higher-order moments are constant.

The autocovariance depends on the scale on which Z is measured, and it is often more convenient and easier to appreciate if it is made dimensionless by converting it to the *autocorrelation*:

$$\rho(\mathbf{h}) = C(\mathbf{h})/C(\mathbf{0}), \tag{4.8}$$

where $C(\mathbf{0})$ is the covariance at lag $\mathbf{0}$, i.e. σ^2.

4.5   INTRINSIC VARIATION AND THE VARIOGRAM

We can represent a stationary random process by the model

$$Z(\mathbf{x}) = \mu + \varepsilon(\mathbf{x}). \tag{4.9}$$

This states simply that the value of Z at $\mathbf{x}$ is the mean of the process plus a random component drawn from a distribution with mean zero and covariance function

$$C(\mathbf{h}) = \mathrm{E}[\varepsilon(\mathbf{x})\varepsilon(\mathbf{x} + \mathbf{h})]. \tag{4.10}$$

Quite the most serious worry and widespread departure from weak stationarity is that the mean appears to change across a region and the variance to increase without bound as the area of interest increases. In these circumstances the covariance cannot be defined. We cannot insert a value for μ in equation (4.7), for example.

Matheron (1965) recognized the problem that this created, and his solution was a major contribution to practical geostatistics. He took the view that, whereas in general the mean might not be constant, it would be so for small $|\mathbf{h}|$ at least, so that the expected differences would be zero:

$$\mathrm{E}[Z(\mathbf{x}) - Z(\mathbf{x} + \mathbf{h})] = 0. \tag{4.11}$$

Further, he replaced the covariances by the variances of differences as measures of spatial relation, which, like the covariance, depended on the lag and not on absolute position. This led to

$$\mathrm{var}[Z(\mathbf{x}) - Z(\mathbf{x} + \mathbf{h})] = \mathrm{E}[\{Z(\mathbf{x}) - Z(\mathbf{x} + \mathbf{h})\}^2]$$
$$= 2\gamma(\mathbf{h}). \tag{4.12}$$

Equations (4.11) and (4.12) constitute Matheron's *intrinsic hypothesis*. This step released practitioners from the constraints of second-order stationarity where the assumptions either did not hold or were doubtful. It opened up a wider field of application. The quantity $\gamma(\mathbf{h})$ is known as the *semivariance* at lag $\mathbf{h}$. The 'semi' evidently refers to the fact that it is half of a variance; it is half the variance of a difference in this instance. It is, nevertheless, the variance per point when the points are considered in pairs, and it had been recognized by Yates (1948). As a function of $\mathbf{h}$ it is the semivariogram, now usually termed simply the *variogram*.

4.5.1   Equivalence with covariance

For second-order stationary processes the variogram and the covariance are equivalent, and from their definitions in equations (4.7) and (4.12) we have

$$\gamma(\mathbf{h}) = C(\mathbf{0}) - C(\mathbf{h}). \tag{4.13}$$

Thus, a graph of the variogram is simply a mirror image of the covariance function about a line or plane parallel to the abscissa. We can also relate the semivariance to the autocorrelation coefficient by combining equations (4.13) and (4.8):

$$\gamma(\mathbf{h}) = \sigma^2 \{1 - \rho(\mathbf{h})\}. \tag{4.14}$$

If a process is intrinsic only there is no equivalence because the covariance function does not exist. The variogram is still valid, nevertheless, and it is its validity in the wider range of circumstances that has made it so much more useful than the covariance. As a consequence it has become the cornerstone of practical geostatistics. For this reason we look at its properties in detail both in the remainder of this chapter and in the following two.

4.5.2 Quasi-stationarity

In practice it often happens that the variogram is of interest only very locally—we shall see this later when we deal with kriging. In these circumstances we can restrict the mean, μ, to that in small neighbourhoods, V, so that equation (4.9) becomes

$$Z(\mathbf{x}) = \mu_V + \varepsilon(\mathbf{x}). \tag{4.15}$$

Provided $\mathbf{h}$ remains within the bounds of V the variogram is unaffected.

4.6 CHARACTERISTICS OF THE SPATIAL CORRELATION FUNCTIONS

We now consider the more important characteristics of the covariance and autocorrelation functions and the variogram. Figure 4.3 illustrates some of these.

Autocorrelation. Like the ordinary product-moment correlation coefficient, the autocorrelation function varies between 1 and -1. From equation (4.8), its value at lag $\mathbf{0}$ is 1.

Symmetry. Because of our assumption of stationarity,

$$
\begin{aligned}
C(\mathbf{h}) &= E[\{Z(\mathbf{x}) - \mu\}\{Z(\mathbf{x} + \mathbf{h}) - \mu\}] \\
&= E[Z(\mathbf{x} - \mathbf{h})Z(\mathbf{x}) - \mu^2] \\
&= E[Z(\mathbf{x})Z(\mathbf{x} - \mathbf{h}) - \mu^2] \\
&= C(-\mathbf{h}). \tag{4.16}
\end{aligned}
$$

In words, the autocovariance is symmetric in space. The same is true of the variogram; i.e. $\gamma(\mathbf{h}) = \gamma(-\mathbf{h})$ for all $\mathbf{h}$. So all three functions are even.

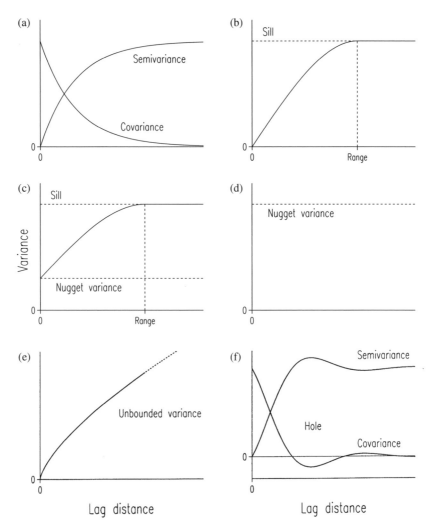

Figure 4.3   Theoretical functions for spatial correlation: (a) typical variogram and equivalent covariance function; (b) bounded variogram showing the sill and range; (c) bounded variogram with a nugget variance; (d) pure nugget variogram; (e) unbounded variogram; (f) variogram and covariance function illustrating the hole effect.

This means that we need consider only the positive lags, and indeed this is the convention. In the graphs of the functions, such as those in Figure 4.3, we show only the right-hand halves of the functions.

Positive semidefiniteness.   The covariance matrix for any number of points is positive semidefinite. That is to say that for a matrix of order n its determinant

$$
\begin{vmatrix}
C(\mathbf{x}_1,\mathbf{x}_1) & C(\mathbf{x}_1,\mathbf{x}_2) & \cdots & C(\mathbf{x}_1,\mathbf{x}_n) \\
C(\mathbf{x}_2,\mathbf{x}_1) & C(\mathbf{x}_2,\mathbf{x}_2) & \cdots & C(\mathbf{x}_2,\mathbf{x}_n) \\
\vdots & \vdots & \cdots & \vdots \\
C(\mathbf{x}_n,\mathbf{x}_1) & C(\mathbf{x}_n,\mathbf{x}_2) & \cdots & C(\mathbf{x}_n,\mathbf{x}_n)
\end{vmatrix}
$$

and all its principal minors are positive or zero. This is necessary because the variance of any linear sum of the random variables,

$$
Y(\mathbf{x}) = \lambda_1 Z(\mathbf{x}_1) + \lambda_2 Z(\mathbf{x}_2) + \cdots + \lambda_n Z(\mathbf{x}_n), \tag{4.17}
$$

must be positive or zero; a variance cannot be negative. The covariance and autocorrelation functions are positive semidefinite. In like manner, the variogram must be negative semidefinite. We shall develop this in Chapter 6, where we shall see that this limits the choice of legitimate mathematical functions to describe the covariance function.

Continuity. As mentioned above, most environmental variables are continuous; the stochastic processes that we believe to represent them are continuous, and so also are the autocovariance functions and variograms of a continuous lag. Crucially, $C(\mathbf{h})$ and $\gamma(\mathbf{h})$ are continuous at $\mathbf{h} = \mathbf{0}$, and if that is so they must be continuous everywhere. So $C(\mathbf{h})$ declines from some positive value, $C(\mathbf{0}) = \sigma^2$, at $\mathbf{0}$ to smaller values at longer lag distances, see Figure 4.3(a). Its mirror image, the variogram increases from 0 at $\mathbf{h} = \mathbf{0}$; i.e. it must pass through the origin if the process is continuous, see Figure 4.3(a)–(b).

If this were not so then we should have a continuous sequence of positions in space, the values at which are not related. It seems impossible, yet in practice data often suggest that a spatial process is discontinuous. It manifests itself most evidently in the sample variogram; the calculated values appear to approach some positive value on the ordinate as the lag distance approaches 0, whereas, at $\mathbf{h} = \mathbf{0}$, $\gamma(\mathbf{0})$ must be 0, Figure 4.3(c). This discrepancy is known as the *nugget variance*. The term arose in gold mining from the notion that gold nuggets occur quite independently of one another at random; they are sparse and certainly not continuous at the working scale. They have a variance that jumps from 0 at lag zero to positive immediately away from the origin, and we can recognize this by defining

$$
\gamma(\mathbf{h}) = \sigma^2 \{1 - \delta(\mathbf{h})\}, \tag{4.18}
$$

where $\delta(\mathbf{h})$ is the Kronecker delta function taking the values 1 when $\mathbf{h} = \mathbf{0}$ and 0 otherwise.

The data themselves differ from their neighbours in irregular steps, large or small, rather than in smooth progression. It seems as though they derive from two or more components, one uncorrelated superimposed on another that is correlated. In other words, we seem to have one source of variation in which contiguous positions in space do take values of Z that are totally unrelated.

Engineers recognize this uncorrelated variation as 'white noise'. They usually express it by its covariance function,

$$C(\mathbf{h}) = \sigma^2 \delta(\mathbf{h}), \tag{4.19}$$

where now $\delta(\mathbf{h})$ is the Dirac function taking the values 0 when $|\mathbf{h}| \neq 0$ and infinity when $|\mathbf{h}| = 0$. Thus for white noise $C(\mathbf{h}) = 0$ for all $|\mathbf{h}| > 0$ and $C(\mathbf{0}) = \infty$. The representation might seem bizarre, but it is the only way that we can describe white noise using covariances. Its equivalent is a 'pure nugget' variogram, Figure 4.3(d).

For properties that vary continuously in space, such as the soil's pH, the concentrations of trace metals, air temperature and rainfall, the apparent nugget variance comprises measurement error plus variation that occurs over distances less than the shortest sampling interval. The latter is usually dominant.

Monotonic increasing. The variograms in Figure 4.3(b)–(c) are monotonically increasing functions, i.e. the variance increases with increasing lag distance. The small $\gamma(\mathbf{h})$ values at short $|\mathbf{h}|$ show that the $Z(\mathbf{x})$ are similar, and that as $|\mathbf{h}|$ increases $Z(\mathbf{x})$ and $Z(\mathbf{x} + \mathbf{h})$ become increasingly dissimilar on average. Looked at from the point of view of correlation, $\rho(\mathbf{h})$ increases as the lag distance shortens, and the process is therefore said to be *autocorrelated* or *spatially dependent*.

Sill and range. The variograms of second-order stationary processes reach upper bounds at which they remain after their initial increases, as in Figure 4.3(b)–(c). The maximum is known as the *sill* variance; it is the *a priori* variance, σ^2, of the process.

A variogram may reach its sill at a finite lag distance, in which case it has a *range*, also known as the *correlation range* since this is the range at which the autocorrelation becomes 0, Figure 4.3(c). This separation marks the limit of spatial dependence. Places further apart than this are spatially independent. Some variograms approach their sills asymptotically, and so they have no strict ranges. For practical purposes their effective ranges are usually taken as the lag distances at which they reach 0.95 times their sills.

Unbounded variogram. If, as in Figure 4.3(e), the variogram increases indefinitely with increasing lag distance then the process is not second-order stationary. It might be intrinsic, but the covariance does not exist.

Hole effect. In some instances the variogram decreases from its maximum to a local minimum and then increases again, Figure 4.3(f). This maximum is equivalent to a minimum in the covariance function, which appears as a 'hole'. This form arises from fairly regular repetition in the process. A variogram that continues to fluctuate with a wave-like form with increasing lag distance signifies greater regularity.

Anisotropy. Spatial variation is not necessarily the same in all directions. If the process is anisotropic, then so is the variogram, as is the covariance function if it exists. Anisotropy may take several forms. The initial gradient may vary. If the variogram has a sill then variation in the gradient will lead to variation in the range, or effective range. If the variation with direction is such that a simple transformation of the spatial coordinates will remove it, then we have *geometric anisotropy* (see Chapter 6).

Different sills, however, mean different amounts of variation in different directions. This is called *zonal anisotropy*, and is the more difficult to imagine.

Drift. In some instances the variogram approaches the origin with a decreasing gradient: it has a concave upwards form. This can arise from local *trends* or *drift*, i.e. smooth change. In these circumstances the expected value, $E[Z(\mathbf{x})]$, is no longer constant, even within small neighbourhoods, but is a function of position. The model for spatial variation becomes

$$Z(\mathbf{x}) = u(\mathbf{x}) + \varepsilon(\mathbf{x}). \tag{4.20}$$

The quantity $u(\mathbf{x})$ is called the 'drift', and it replaces the means in equations (4.9) and (4.15). The assumption of second-order stationarity does not hold, nor even does the intrinsic hypothesis: the raw semivariances no longer estimate the expected squared differences between the residuals at two places. The residuals are given by

$$\varepsilon(\mathbf{x}) = Z(\mathbf{x}) - u(\mathbf{x}), \tag{4.21}$$

and they constitute a random process.

A more general description of non-stationarity is as an *intrinsic random function of order k*:

$$Z(\mathbf{x}) = Z_k(\mathbf{x}) + u(\mathbf{x}). \tag{4.22}$$

4.7 WHICH VARIOGRAM?

The variogram (and covariance function) as treated above is a function of an underlying stochastic process. We may regard it as the theoretical variogram. It may be thought of as the average of the variograms from all possible realizations of the process. Following Matheron (1965), we need to distinguish it from two others, namely the regional and the experimental.

The *regional variogram* is the variogram of the particular realization in a finite region, R. It is the one that you might compute if you had complete information of the region, as, for example, from the simulated fields in Figures 6.4, 6.5, 6.7 and 6.8, and from many digital images (see Muñoz-Pardo, 1987).

Because the realization is bounded we cannot assume that the regional variogram fully represents the ensemble. A process might be second-order stationary, but appear unbounded in a small region, especially if the distance across the region is smaller than the correlation range. The regional variogram is called the non-ergodic variogram by some workers, e.g. Brus and de Gruijter (1994), for this reason. It is more or less accessible, depending on the effort we are prepared to devote to sampling the realization, and this leads us to the third variogram, below.

The *experimental variogram* is a variogram computed from data, $z(\mathbf{x}_i)$, $i = 1, 2, \ldots$, which constitute a sample from the region. It is also called the *sample variogram*. We describe it in Chapter 5. It necessarily applies to an actual realization, and it estimates the regional variogram for that realization. It is usually the only variogram that we know, and any inference from it requires modelling, as described in Chapter 6.

4.8 SUPPORT AND KRIGE'S RELATION

Spatial dependence within a finite region has both theoretical and practical consequences, which we now explore.

The variance of $Z(\mathbf{x})$ within a region R of area $|R|$ is the double integral of the variogram:

$$\sigma_R^2 = \overline{\gamma}(R, R) = \frac{1}{|R|^2} \int_R \int_R \gamma(\mathbf{x} - \mathbf{x}') \, \mathrm{d}\mathbf{x}\mathrm{d}\mathbf{x}', \qquad (4.23)$$

where $\mathbf{x}$ and $\mathbf{x}'$ sweep independently over R. In geostatistics this variance is called the *dispersion variance* of $Z(\mathbf{x})$ in R. Unless the variogram is all nugget the dispersion variance for a finite R is less than the *a priori* variance of the process, assuming second-order stationarity. Figure 4.4 shows the relation between the two for a one-dimensional process; in it the shaded areas are equal. Evidently, as R is made smaller σ_R^2 diminishes, until in the limit we are left with a point, at which σ_R^2 disappears.

The region R (see Figure 4.5) limits the extent of a realization. At the small end of our spatial scale we encounter another limit. Measurements must be made on finite volumes, whether of samples taken into the laboratory or the surroundings of instruments placed in the field. The volume, with its particular size, shape and orientation, is known as the *support* of the sample. The supports have finite cross-sectional areas in $\mathbb{R}^2$, and they are effectively tiny but finite regions, each with its own dispersion variance. If we denote them by b, each with area $|b|$ (see Figure 4.5), then their

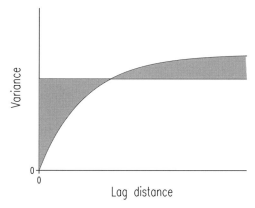

Figure 4.4 Relation between variogram and the dispersion variance (the horizontal line) in a finite region, R.

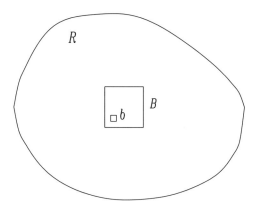

Figure 4.5 Krige's relation for a region, R, a block, B, and a small support, b.

dispersion variances are given by the analogue of equation (4.23):

$$\sigma_b^2 = \overline{\gamma}(b, b) = \frac{1}{|b|^2} \int_b \int_b \gamma(\mathbf{x} - \mathbf{x}') \, \mathrm{d}\mathbf{x} \mathrm{d}\mathbf{x}'. \tag{4.24}$$

One practical consequence of this is that the support of the sample sets a minimum to the resolution of the spatial variation that can be detected and measured by that sample: engineers will understand this as 'band-limited' measurement.

In many applications we are interested in blocks, B, of intermediate size, $|B|$ (see Figure 4.5). They may be mining blocks, plots in an experiment, or fields on a farm, for example. They too will have dispersion variances, σ_B^2, defined in a way analogous to σ_R^2 and σ_b^2, and with intermediate values. We now relate the three.

Consider first the supports b. Though tiny, they have finite size, and so in a finite region they are finite in number if they do not overlap. If there are n_R^b of them with values z_i^b, $i = 1, 2, \ldots, n_R^b$, then their variance in R is

$$s^2(b \in R) = \frac{1}{n_R^b} \sum_{i=1}^{n_R^b} \{\bar{z}_R - z_i^b\}^2, \qquad (4.25)$$

where $\bar{z}_R$ is the mean of the z_i^b. In like manner their variance in a block B with mean $\bar{z}_B$ is

$$s^2(b \in B) = \frac{1}{n_B^b} \sum_{i=1}^{n_B^b} \{\bar{z}_B - z_i^b\}^2, \qquad (4.26)$$

which can be averaged over all $B \in R$ to give $\bar{s}^2(b \in B)$. Finally we consider the blocks, B, themselves. Their variance in R is

$$s^2(B \in R) = \frac{1}{n_R^B} \sum_{j=1}^{n_R^B} \{\bar{z}_R - \bar{z}_j^B\}^2, \qquad (4.27)$$

where $\bar{z}_j^B$ is the mean of Z in the jth block.

For any finite region that is divided in the above way into blocks, which in turn are further subdivided, whether into small supports or smaller blocks, the dispersion variance is partitioned quite simply as

$$s^2(b \in R) = \bar{s}^2(b \in B) + s^2(B \in R). \qquad (4.28)$$

In words, the dispersion variance of Z of supports b in region R is the sum of the variance of the supports with blocks B plus the variance of the blocks within R. This is *Krige's relation*. It is strictly analogous to the partition of the total variance into within and between classes in the simple one-way analysis of variance.

The expectations of the dispersion variances are all readily obtained from the variogram by

$$\sigma^2(b \in R) = \bar{y}(R, R) - \bar{y}(b, b),$$
$$\sigma^2(B \in R) = \bar{y}(R, R) - \bar{y}(B, B), \qquad (4.29)$$
$$\sigma^2(b \in B) = \bar{y}(B, B) - \bar{y}(b, b),$$

and so Krige's relation applies to them equally.

4.8.1　Regularization

Another consequence of the finite sample support is that the variogram in practice is a function of the support. The larger the support is the more

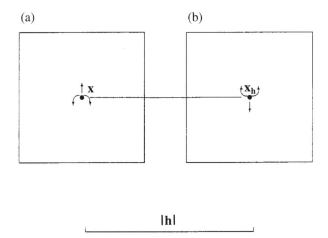

Figure 4.6   The block-to-block integration of the variogram.

variation each measurement encompasses, and the less there is in the intervening space to appear in the variogram. This inevitably diminishes the sill or gradient and tends to make the variogram concave upwards near to the origin. It is a physical regularization, the statistical aspects of which we describe below. Results should always refer specifically to the particular support, which should therefore remain the same throughout any one investigation.

The variogram on one support can be related, at least theoretically, to that on another. The semivariance for two supports $b(\mathbf{x})$ and $b(\mathbf{x} + \mathbf{h})$, the centroids of which are $\mathbf{h}$ apart, is

$$\gamma_b(\mathbf{h}) = E[\{Z_b(\mathbf{x}) - Z_b(\mathbf{x} + \mathbf{h})\}^2], \qquad (4.30)$$

where $Z_b(\mathbf{x})$ and $Z_b(\mathbf{x} + \mathbf{h})$ are the integrals of $Z(\mathbf{x})$ over the supports b. This is composed of two parts, the average squared difference between the points in one support and those in the other less the dispersion variance within supports. The first is given by

$$\overline{\gamma}(b, b_{\mathbf{h}}) = \frac{1}{|b|^2} \int_b \int_b \gamma(\mathbf{x} - \mathbf{x}_{\mathbf{h}}) \, d\mathbf{x} d\mathbf{x}_{\mathbf{h}}, \qquad (4.31)$$

where $\mathbf{x}$ sweeps one support and $\mathbf{x}_{\mathbf{h}}$ sweeps the other independently, as in Figure 4.6. The second is the integral of the variogram within the support b:

$$\overline{\gamma}(b, b) = \frac{1}{|b|^2} \int_b \int_b \gamma(\mathbf{x} - \mathbf{x}') \, d\mathbf{x} d\mathbf{x}', \qquad (4.32)$$

where $\mathbf{x}$ and $\mathbf{x}'$ sweep b independently, as illustrated in Chapter 8 (Figure 8.1). The variogram on the new supports thus becomes

$$\gamma_b(\mathbf{h}) = \overline{\gamma}(b, b_{\mathbf{h}}) - \overline{\gamma}(b, b). \qquad (4.33)$$

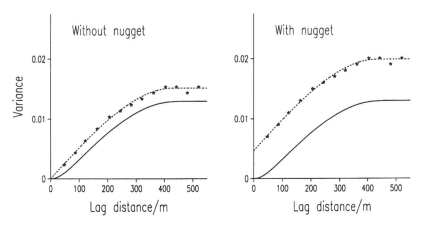

Figure 4.7 Punctual (dashed line) and regularized (solid line) variograms on a support of 80 m × 80 m.

If $|\mathbf{h}|$ is large relative to the distances across the support then $\overline{\gamma}(b, b_\mathbf{h})$ is approximately the punctual semivariance at lag $\mathbf{h}$, and

$$\gamma_b(\mathbf{h}) \approx \gamma(\mathbf{h}) - \overline{\gamma}(b, b). \tag{4.34}$$

So when $|\mathbf{h}| \gg \sqrt{\text{area of } b}$ the regularized variogram is derived from the punctual one simply by subtracting the dispersion variance of the support.

This procedure in which the variogram for one support is obtained from that of a smaller support is known as *regularization*. The result is shown in Figure 4.7: the sill has diminished, the nugget variance has disappeared and the approach of the variogram at the origin is somewhat concave upwards. It is especially important when bulking samples, because (a) the supports can be large, and (b) because if the variogram is known for very small supports on which the variable has been measured then that for samples bulked over larger areas, the regularized variogram, can be determined from it and surveys planned with greater efficiency.

5

Estimating the variogram

The variogram is the cornerstone of geostatistics, and it is therefore vital to estimate it and model it correctly. This chapter concerns its estimation and several other related matters which influence the reliability of the result and how we interpret it. It includes the computing equations and robust estimators, the statistical distribution of the data, trend, anisotropy, the effect of sample size on the confidence we can have in a variogram, and sample design. Our aim is not only to estimate the regional variogram reliably but also to show how to use limited resources wisely.

5.1 ESTIMATING SEMIVARIANCES AND COVARIANCES

5.1.1 The variogram cloud

For any set of data we can compute the semivariance for each pair of points, $\mathbf{x}_i$ and $\mathbf{x}_j$ individually as

$$y(\mathbf{x}_i, \mathbf{x}_j) = \tfrac{1}{2}\{z(\mathbf{x}_i) - z(\mathbf{x}_j)\}^2. \tag{5.1}$$

These values can then be plotted against the lag distance as a scatter diagram, called the 'variogram cloud' by Chauvet (1982). Figure 5.1 shows the variogram cloud for $\log_{10} K$ at Broom's Barn Farm to a lag of 600 m. It contains all of the information on the spatial relations in the data to that lag. In principle, we could fit a model to it to represent the regional variogram, but in practice it is almost impossible to judge from it if there is any spatial correlation present, what form it might have, and how we could model it. A more sensible approach is to average the semivariances for each of a few lags and then examine the result. Nevertheless, the variogram cloud shows the spread of values at the different lags, and it might enable us to detect outliers or anomalies. The tighter this distribution is, the stronger is the spatial continuity in the data.

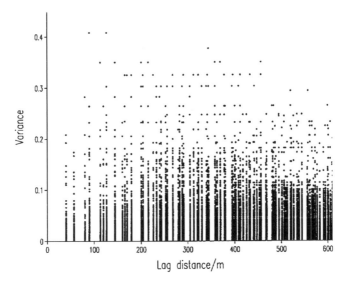

Figure 5.1 A variogram cloud.

5.1.2 Average semivariances

If we recall the definition of the semivariance from equation (4.12) as

$$\gamma(\mathbf{h}) = \tfrac{1}{2}\mathrm{E}[\{Z(\mathbf{x}) - Z(\mathbf{x} + \mathbf{h})\}^2] \tag{5.2}$$

then its estimator is

$$\hat{\gamma}(\mathbf{h}) = \tfrac{1}{2}\mathrm{mean}[\{z(\mathbf{x}) - z(\mathbf{x} + \mathbf{h})\}^2], \tag{5.3}$$

where the $z(\mathbf{x})$ and $z(\mathbf{x} + \mathbf{h})$ represent actual values of Z at places separated by $\mathbf{h}$. For a set of data $z(\mathbf{x}_i)$, $i = 1, 2, \ldots$, we can compute

$$\hat{\gamma}(\mathbf{h}) = \frac{1}{2m(\mathbf{h})} \sum_{i=1}^{m(\mathbf{h})} \{z(\mathbf{x}_i) - z(\mathbf{x}_i + \mathbf{h})\}^2, \tag{5.4}$$

where $m(\mathbf{h})$ is the number of pairs of data points separated by the particular lag vector $\mathbf{h}$. This is the usual computing formula, though the way that it is implemented as an algorithm depends on the configuration of the data, and we consider the possibilities below.

Regular sampling in one dimension

For regular sampling in one dimension along transects and down boreholes we can denote the data by $z_i = z(\mathbf{x}_i)$, $i = 1, 2, \ldots, N$. The lag becomes a scalar, $h = |\mathbf{h}|$, for which $\hat{\gamma}$ can be computed only at integral

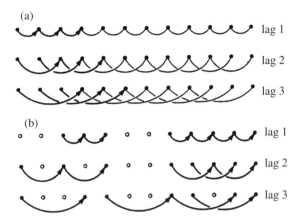

Figure 5.2 Comparisons for computing a variogram from regular sampling on a transect: (a) without missing values; (b) with missing values.

multiples of the sampling interval. The semivariance is then computed as

$$\hat{\gamma}(h) = \frac{1}{2(N-h)} \sum_{i=1}^{N-h} \{z_i - z_{i+h}\}^2. \tag{5.5}$$

Figure 5.2(a) shows the situation. First, the squared differences between neighbouring pairs of values, z_1 and z_2, z_2 and z_3, and so on, i.e. for $h = 1$, are determined for each position and averaged. All of the observations at lag interval h are used twice except for those at the ends of the transect, and so there are $N - 1$ comparisons. If there are missing values at some locations, as in Figure 5.2(b), then there will be fewer comparisons, and the divisor is diminished accordingly. By increasing h to 2 the comparisons are then z_1 with z_3, z_2 with z_4, etc., and we can repeat the procedure for $h = 3, 4, \ldots$. The result is a set of semivariances $\hat{\gamma}(1), \hat{\gamma}(2), \hat{\gamma}(3), \ldots,$ that is ordered as a function of h. It is a one-dimensional *sample variogram* or *experimental variogram*, and we can plot $\hat{\gamma}(h)$ against h as in Figure 5.3.

Irregular sampling in one dimension

If data are irregularly scattered then the average semivariance for any particular lag can be derived only by grouping the individual lag distances between pairs of points. Otherwise we have individual semivariances as in the variogram cloud. Typically the averaging is done by choosing a set of lags, $h_j, j = 1, 2, \ldots$, at arbitrary constant increments d, and then associating with each h_j a class of width d, bounded by $h_j - d/2$ and $h_j + d/2$. Each pair of points separated by $h_j \pm d/2$ is used to estimate $\gamma(h_j)$. In this way each comparison contributes to one and only one estimate. Sometimes there are more comparisons at the shorter lags, especially where

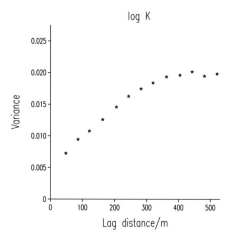

Figure 5.3   A sample variogram (of $\log_{10} K$ at Broom's Barn Farm).

sampling has been nested (see below), and then it can be advantageous to increase the increments, and with them d, as h increases.

The lag increments can affect the resulting variogram, and so d should be chosen with care. If the increment is small then there might be too few comparisons at each lag, leading to semivariances that are estimated crudely and an experimental variogram that appears erratic. If, on the other hand, d is large then there are likely to be few estimates and detail is lost by unnecessary smoothing. The best compromise will depend on the number of data, the evenness of the sampling and the form of the underlying variogram. A useful starting point is to use the average separation between nearest neighbours as the interval.

Sampling on transects to represent variation in two dimensions

Sometimes investigators sample regularly along transects to explore variation in two dimensions and, in particular, to identify and estimate anisotropy, i.e. directional differences. The computational procedure is the same as for the regular one-dimensional sampling, and equation (5.5) produces a separate set of estimates for each transect. These need to be seen together as a whole and not as separate variograms. The variogram in two dimensions is itself two-dimensional, and the ordered sets of semivariances computed from transects are effectively samples of sections through the two-dimensional function. To identify and estimate anisotropy transects must be aligned in at least three directions. If the directional variogram appears to have markedly different gradients or ranges in the different directions then it is likely that the underlying variation is anisotropic, and it should be modelled accordingly (see Chapter 6). If the variation seems isotropic, i.e. there seem to be no directional differences, then the separate estimates can be averaged over all directions to give the isotropic variogram where the vector **h** can be replaced by the scalar $h = |\mathbf{h}|$.

Regular sampling in two dimensions

For data recorded at regular intervals on a rectangular grid the above formula (5.5), for one dimension, is readily extended. If the grid has m rows and n columns then we compute

$$\hat{\gamma}(p,q) = \frac{1}{2(m-p)(n-q)} \sum_{i=1}^{m-p} \sum_{j=1}^{n-q} \{z(i,j) - z(i+p,j+q)\}^2,$$

$$\hat{\gamma}(p,-q) = \frac{1}{2(m-p)(n-q)} \sum_{i=1}^{m-p} \sum_{j=q+1}^{n-q} \{z(i,j) - z(i+p,j-q)\}^2,$$

(5.6)

where p and q are the lags along the rows and down the columns of the grid, respectively. In general, the lag is simply the grid interval. These equations enable half the variogram to be computed for lags from $-q$ to q and from 0 to p. The variogram is symmetrical about its centre, and the full set of semivariances is obtained by computing

$$\hat{\gamma}(-p,q) = \hat{\gamma}(p,-q),$$
$$\hat{\gamma}(-p,-q) = \hat{\gamma}(p,q).$$

(5.7)

The procedure can be envisaged as moving the grid over itself to the right and up or down to new positions, as in Figure 5.4, and making the comparisons between the values at the points that coincide. In Figure 5.4(a), the grid has been moved to the right by three units, i.e. $p = 3$ and $q = 0$, as represented by the horizontal line. In Figure 5.4(b) the grid has been moved down one unit in addition, so that now $q = -1$; the horizontal and vertical shifts are shown in the triangle, with its hypotenuse showing the resultant.

Figure 5.4 also shows that as p and q are increased so the number of coincident points diminishes rapidly from the original 55. As a consequence, the semivariances become less and less well estimated, a matter to which we return below.

Where data are missing the quantities $(m-p)(n-q)$ in the denominators of equation (5.6) must be replaced by the actual numbers of comparisons.

Irregular sampling in two dimensions

Survey data in two dimensions are often unevenly distributed. Each pair of observations is separated by a potentially unique lag in both distance and direction. To obtain averages containing directional information we must group the separations by direction as well as by distance. Figure 5.5

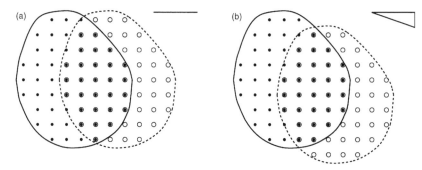

Figure 5.4   Computing a two-dimensional variogram from a regular grid of data by sliding the grid over itself (a) by three units to the right, and (b) by one unit down in addition. The resultant lag is given by the hypotenuse of the triangle.

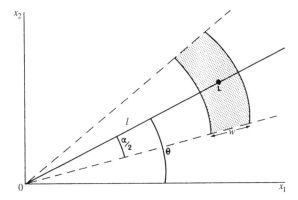

Figure 5.5   The geometry for discretizing the lag by distance and direction where sampling has been irregular in two dimensions.

shows the geometry of the grouping. We choose a lag interval, the multiples of which will form a regular progression of nominal lag distances as in the one-dimensional case. We then choose a range in distance, w in Figure 5.5, usually equal to the lag interval. The nominal lag distance is represented by the line OL of length l. We also choose a set of directions, one of which is shown as θ in Figure 5.5, and a range in direction, α, such that $\alpha = \pi/n$, where n is the number of directions, and θ progresses in steps of α from 0 to $\pi/(n-1)$. For example, if we choose four directions ($n = 4$) then a sensible progression for θ would be 0, $\pi/4$, $\pi/2$, $3\pi/4$, i.e. 0, 45, 90 and 135 degrees, with $\alpha = \pi/4$ (45°). This ensures complete coverage and no overlap between the different directions. For six directions α would be 30°. Then, for a point $\mathbf{x}_i$ at O with a second point $\mathbf{x}_i + \mathbf{h}$ within the stippled zone, $\{z(\mathbf{x}_i) - z(\mathbf{x}_i + \mathbf{h})\}^2$ contributes to $\hat{y}(\mathbf{h}) = \hat{y}(l, \theta)$. When all comparisons have been made the experimental variogram will consist of the set of averages for the nominal lags in both distance and direction. We can extend this further by computing the average experimental variogram over all directions (omnidirectional) by setting $\alpha = \pi$ (180°). Appendix B gives the Genstat instructions for computing directional and omnidirectional variograms.

Exploring and displaying anisotropy

So far we have concentrated on explaining the computation in one and two dimensions, but there is also the matter of representing the results of the two spatial dimensions on a plane, and of exploring differences in the variation in two dimensions.

Where data are on a rectangular grid the semivariances along the rows and columns and those on the principal diagonals can be plotted separately, bearing in mind that the lag intervals will not be the same in all four directions. No directional information is lost, and the results can then be examined for directional differences. Where data are irregularly scattered and we have to group the angular separations we inevitably lose some of the directional information. The wider is α the more information we lose, until when $\alpha = \pi$ (180°) all is lost. Choosing α is therefore a compromise between a stable estimate based on many comparisons over a wide angle that will underestimate variance in the direction of the maximum and overestimate that in the direction of minimum and one that is subject to large error but which gets closer to the true values in the directions of maximum and minimum. At the outset a reasonable rule of thumb is to let $\alpha = \pi/4$. If this appears to reveal anisotropy then try reducing α until the resulting variogram becomes too erratic. The larger is α, the more the anisotropy ratio will be underestimated when models are fitted (see Chapter 6). If the variation is isotropic the vector $\mathbf{h}$ can be replaced by the scalar $h = |\mathbf{h}|$ in distance only, and the general computing formula (5.4) can be used. In this case we set $\alpha = \pi$ to compute the omnidirectional variogram.

Whereas it is easy to draw and comprehend a graph of the experimental variogram for either one-dimensional data or one averaged over all directions in two dimensions, it is much less so for the two-dimensional experimental variogram. One simple way is to plot the values with a unique symbol for each direction on the same pair of axes (Figure 5.6). Alternatively, some kind of statistical surface can be fitted to the two-dimensional variogram to represent it as an isarithmic chart or perspective diagram (Figures 5.7 and 5.8). When the variogram has been modelled, this surface can be that of the model. The ideal solution would be to draw it as a stereogram.

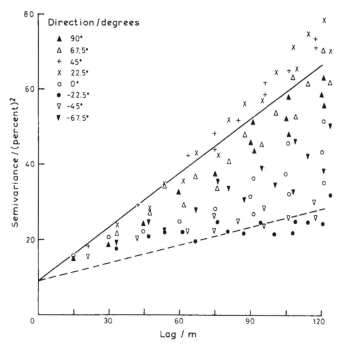

Figure 5.6   A two-dimensional variogram with a distinct symbol for each of eight directions. The oblique lines are the bounds of the fitted model (see Chapter 6).

5.1.3   The experimental covariance function

All of the above considerations also apply to the estimation of spatial covariances, and the equations are analogous. Remember, however, that the covariance requires stationarity of the mean and the variance of the underlying process. The general computing formula for the experimental covariance at lag $\mathbf{h}$, the analogue of equation (5.4), is

$$\hat{C}(\mathbf{h}) = \frac{1}{m(\mathbf{h})} \sum_{i=1}^{m(\mathbf{h})} \{z(\mathbf{x}_i)z(\mathbf{x}_i + \mathbf{h})\} - \bar{z}^2, \tag{5.8}$$

where $\bar{z}$ is the mean of all the data. The analogous correlation function, the sample correlogram, is readily derived from $\hat{C}(\mathbf{h})$ by

$$\hat{\rho}(\mathbf{h}) = \frac{\hat{C}(\mathbf{h})}{s^2}, \tag{5.9}$$

where s^2 is the variance of the data.

If $Z(\mathbf{x})$ is second-order stationary then $\hat{C}(\mathbf{h}) \approx \hat{C}(\mathbf{0}) - \hat{y}(\mathbf{h})$ for all $\mathbf{h}$. If there is trend in the variation then $\hat{C}(\mathbf{0}) - \hat{y}(\mathbf{h})$ will tend to be larger than $\hat{C}(\mathbf{h})$ computed by equation (5.8). This tendency can be counteracted by

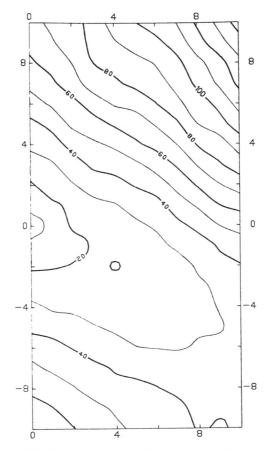

Figure 5.7 An isarithmic chart of a two-dimensional variogram. The origin is in the middle of the left-hand side.

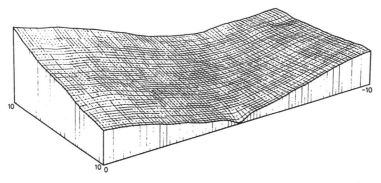

Figure 5.8 A perspective diagram of a two-dimensional variogram. The origin is in the middle at the front.

replacing the regional mean $\bar{z}$ by two distinct means, one the mean of the $z(\mathbf{x}_i)$, $i = 1, 2, \ldots$, say $\bar{z}_1$, and the other the mean of the $z(\mathbf{x}_i + \mathbf{h})$, $\bar{z}_2$, and computing

$$\hat{C}(\mathbf{h}) = \frac{1}{m(\mathbf{h})} \sum_{i=1}^{m(\mathbf{h})} \{z(\mathbf{x}_i) - \bar{z}_1\}\{z(\mathbf{x}_i + \mathbf{h}) - \bar{z}_2\}. \tag{5.10}$$

This measure of the covariance corresponds with that often used in statistics. It does not assume implicitly that $\bar{z}_1$ is the same as $\bar{z}_2$. Several spatial analysts, e.g. Deutsch and Journel (1992) and Isaaks and Srivastava (1989), use this formula as a matter of course. They call the quantities $\bar{z}_1$ and $\bar{z}_2$ the means of the 'heads' and of the 'tails', respectively.

Similarly, the autocorrelation coefficients can be estimated by

$$\hat{\rho}(\mathbf{h}) = \frac{\hat{C}(\mathbf{h})}{s_1 s_2}, \tag{5.11}$$

where s_1 and s_2 are the standard deviations of the heads and tails. These formulae are used by time-series analysts, but Yule and Kendall (1950) warn against equation (5.11) if you have rather few data.

5.1.4   Drift and trend

Variation in $Z(\mathbf{x})$ might contain a systematic component in addition to the random one (see Chapter 4). Equation (4.20) expressed this by

$$Z(\mathbf{x}) = u(\mathbf{x}) + \varepsilon(\mathbf{x}), \tag{5.12}$$

in which $u(\mathbf{x})$ is the drift and is not constant. The expression for the drift replaces the local mean in equation (4.15), and the assumptions of the random function model no longer hold. In these circumstances $\{Z(\mathbf{x}) - Z(\mathbf{x} + \mathbf{h})\}^2$ does not equal $\{\varepsilon(\mathbf{x}) - \varepsilon(\mathbf{x} + \mathbf{h})\}^2$, and the raw semivariances computed by equation (5.4) will be biased estimates of $\gamma(\mathbf{h})$, the variogram of $\varepsilon(\mathbf{x})$.

The variogram describing the random variation is that of the residuals,

$$\varepsilon(\mathbf{x}) = Z(\mathbf{x}) - u(\mathbf{x}). \tag{5.13}$$

To estimate $\gamma(\mathbf{h})$ in this way we need to be able to separate $u(\mathbf{x})$ from $\varepsilon(\mathbf{x})$. Unfortunately we know neither; all we have are the $z(\mathbf{x}_i)$, $i = 1, 2, \ldots, N$. There are several ways out of the impasse, and we look at them briefly.

Long-range trend

If $u(\mathbf{x})$ is long-range trend, e.g. the depth of the water table beneath a slope or the height of a bedding plane in a region of gently folded sediments, then one can postulate a simple mathematical form for it, fit the form by regression to the $z(\mathbf{x}_i)$, and compute the residuals as the $u(\mathbf{x}_i)$ and from them their variogram. We mentioned this in the previous chapter.

The simplest model for non-stationary trend in two dimensions is an inclined plane. Its equation is

$$u(\mathbf{x}) = u(x_1, x_2) = b_0 + b_1 x_1 + b_2 x_2, \qquad (5.14)$$

in which x_1 and x_2 are the spatial coordinates. This is a simple regression of the values on the spatial coordinates.

A quadratic function or second-order polynomial is a slightly more complex expression of the trend, and its equation is

$$u(\mathbf{x}) = u(x_1, x_2) = b_0 + b_1 x_1 + b_2 x_2 + b_3 x_1^2 + b_4 x_2^2 + b_5 x_1 x_2. \quad (5.15)$$

More complex still is the cubic function or third-order polynomial with equation

$$u(\mathbf{x}) = u(x_1, x_2) = b_0 + b_1 x_1 + b_2 x_2 + b_3 x_1^2 + b_4 x_2^2 + b_5 x_1 x_2$$
$$+ b_6 x_1^3 + b_7 x_2^3 + b_8 x_1^2 x_2 + b_9 x_1 x_2^2. \qquad (5.16)$$

We have found that this is the maximum order of polynomial needed to express trend, although more complex functions including cyclic terms are used in trend surface analysis.

Moffat *et al.* (1986) used this procedure to remove the long-range regional dip, which was well known, to identify the short-range flexures in the Cenomanian rocks beneath the Chiltern Hills in south-east England and to map them.

Short-range drift

If $u(\mathbf{x})$ appears to vary smoothly over short distances then another approach must be tried. Again we can assume a simple functional form for it,

$$u(\mathbf{x}) = \sum_{k=0}^{K} a_k f_k(\mathbf{x}), \qquad (5.17)$$

in which a_k, $k = 0, 1, \ldots, K$, are unknown coefficients, and the $f_k(\mathbf{x})$ are known functions of $\mathbf{x}$ of our choosing. We also assume a theoretical variogram, and we use it to evaluate the coefficients a_k in equation (5.17). This gives values for $u(\mathbf{x})$. Using these estimates of the drift, we obtain residuals between them and the data and compute an experimental variogram of the residuals. This should match the theoretical variogram.

Deriving the coefficients a_k in equation (5.17) requires a full structural analysis and is lengthy; Olea (1975) provides all the details. In practice, we can keep the task fairly simple by sampling initially on a regular grid, by analysing one direction at a time, and by restricting the neighbourhood over which we compute the variogram, as in the stationary case. In these circumstances the expression for the drift need never be more complex than a second-order polynomial, and the variogram can be well approximated by a straight line,

$$\gamma(h) = wh,\tag{5.18}$$

usually without a nugget.

Equation (5.17) for the drift in one dimension can thus take one of the following forms:

$$u(x) = a_0,$$
$$u(x) = a_0 + a_1,\tag{5.19}$$
$$u(x) = a_0 + a_1 + a_2,$$

where x relates to an origin within the neighbourhood. The expressions are combined with a linear variogram that can be changed by lengthening or shortening the neighbourhood over which it applies.

The constant a_0 cannot be determined, but that does not matter because the variogram of the residuals does not depend on it and so it is not needed. Olea (1975) shows that with regular sampling the other coefficients are given by

$$a_1 = \frac{z(x_n) - z(x_1)}{d(n-1)},\tag{5.20}$$

for linear drift, and for quadratic drift they are

$$a_1 = \frac{z(x_n) - z(x_1)}{d(n-1)} - (n-1)a_2 d,$$
$$a_2 = -\frac{3\{2\bar{z} - z(x_n) - z(x_1)\}}{d(n-2)(n-1)d^2}.\tag{5.21}$$

In these equations x_1 and x_n are the extreme sampling positions in the neighbourhood, d is the sampling interval, and $\bar{z}$ is the mean of Z in the neighbourhood.

When the residuals from the drift have been calculated their semivariances are biased estimates of the true residuals. The bias can be estimated by assuming that the variogram is linear and that its slope is well estimated by the gradient of the variogram at the origin, effectively the gradient over the first lag interval.

For a linear drift the gradient is

$$w = \frac{n-1}{n-2}\gamma^*(1),\tag{5.22}$$

where $\gamma^*(1)$ is the estimated semivariance at lag 1. The bias at lag h which we denote $p(h)$, is then

$$p(h) = \frac{wh^2}{d(n-1)},\tag{5.23}$$

and the semivariances of the true residuals are

$$y(h) = y^*(h) + p(h).$$ (5.24)

For quadratic drift the gradient of the true variogram is estimated from $y^*(1)$ by

$$w = \frac{n-1}{n-3} y^*(1),$$ (5.25)

and the full semivariances are given by

$$y(h) = y^*(h) + \frac{wh^2\{d^2(2n^2 - 2n - 1) - 2dnh + h^2\}}{(n-2)(n-1)nd^3}.$$ (5.26)

So, to proceed, first choose a combination of drift power and size of neighbourhood, and do the analysis as above. If the experimental variogram matches the theoretical one then the combination is regarded as a good description of the variation. If, on the other hand, the match is poor then you should change either the expression for the drift or the size of the neighbourhood or both and repeat the analysis. It is not always easy to judge whether a particular combination is satisfactory, and even if a match seems good there might be better combinations. So you repeat the analysis with successively larger neighbourhoods first for linear drift and then for quadratic drifts, and compare the results. The best fit is chosen by inspecting graphs of the results.

The following example from Webster and Burgess (1980) illustrates the approach. The data are of electrical resistivity (in ohm metre) in the topsoil over an archaeological site in south-east England measured by inserting probes into the soil at 1 m intervals on a regular square grid. Figure 5.9 shows the raw variogram in four directions. Figures 5.10 and 5.11 show some of the results of the structural analysis. In the graphs the circles represent the assumed linear variogram, and stars are the actual semivariances of the residuals.

The agreement between assumed and actual variograms is good for both linear and quadratic drifts when the neighbourhood is restricted to five terms. With nine terms in the neighbourhood agreement is still good for the quadratic drift but somewhat less so for linear drift. On increasing the size of the neighbourhood the agreement steadily deteriorates (Webster and Burgess, 1980).

The above approach, combining equations (5.17) and either (5.24) or (5.26), is the basis of *universal kriging* described in Chapter 8. The potential bias in universal kriging from estimating the variogram of the residuals led Matheron (1973) to suggest an alternative approach: intrinsic random functions of order k (IRFk), where positive values of k represent a series of polynomials. The basic aim is to filter the values of Z linearly. An IRF0 simply represents an intrinsic stationary process. Matheron showed

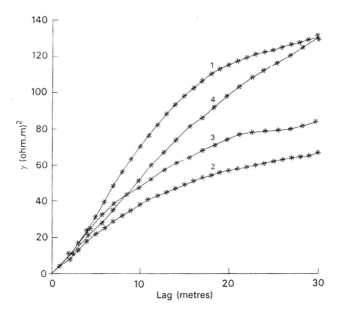

Figure 5.9   The raw variogram of resistivity at Bekesbourne.

that an IRFk has a generalized covariance function. This is essentially the covariance of the differences between the residuals from the trend for a given separation **h**. Readers who wish to pursue this are recommended to read Cressie (1993) and Wackernagel (1995).

5.2   RELIABILITY OF THE EXPERIMENTAL VARIOGRAM

Apart from the matter of anisotropy, the above equations provide asymptotically unbiased estimates of $y(\mathbf{h})$ for Z in the region of interest, R. However, the variogram obtained will fluctuate more or less, and so will its reliability, and we now look at factors that affect these.

5.2.1   Statistical distribution

The variogram is sensitive to outliers and to extreme values in general. It will be evident from equations (5.5) and (5.6), for example, that each observed $z(\mathbf{x})$ can contribute to several estimates of $y(\mathbf{h})$. So one exceptionally large or exceptionally small $z(\mathbf{x})$ will tend to swell $\hat{y}(\mathbf{h})$ wherever it is compared with other values. The result is to inflate the average.

However, the effect is not uniform. If the extreme is near the margin of the region then it will contribute to fewer comparisons than if it is near the centre. The end point on a regular transect, for example, contributes

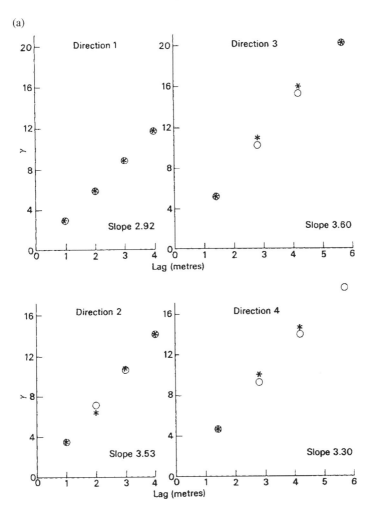

Figure 5.10 Variograms of a structural analysis of the Bekesbourne data for a linear drift over neighbourhoods of (a) five terms (four lag intervals).

to the average just once for each lag, whereas points near the middle contribute many times. If data are unevenly scattered then the relative contributions of extreme values are even less predictable. The result is that the experimental variogram is not inflated equally over its range, and this can add to its erratic appearance.

Exceptional contributions to the semivariances can be identified by drawing an **h**-scattergram for each lag, **h**. An **h**-scattergram is a graph in which the $z(\mathbf{x})$ are plotted against the $z(\mathbf{x} + \mathbf{h})$ with which they are compared in computing $\hat{y}(\mathbf{h})$. In general the plotted points appear as more or

(b)

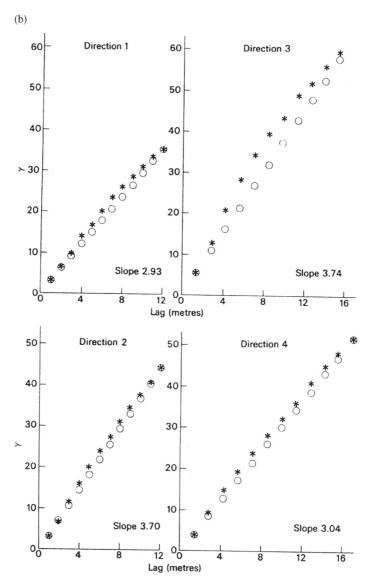

Figure 5.10 (*Cont.*) Variograms of a structural analysis of the Bekesbourne data for a linear drift over neighbourhoods of (b) nine terms (eight lag intervals).

less inflated clusters, as in the usual kind of scatter diagram. The closer the points lie to the diagonal line with gradient 1, the stronger is the correlation, i.e. the larger is $\hat{\rho}(\mathbf{h})$ and the smaller is $\hat{y}(\mathbf{h})$.

Figure 5.12 shows four such graphs for lag distances 40 m, 80 m, 120 m and 160 m; these are the first four lag distances in Figure 5.3, the vario-

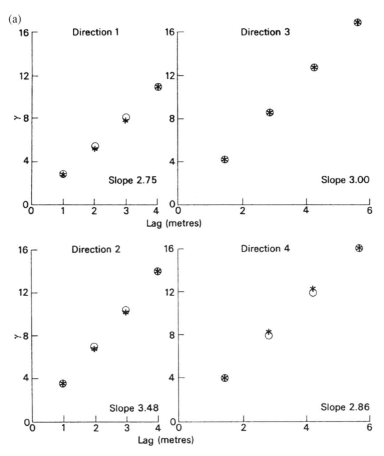

Figure 5.11 Variograms of a structural analysis of the Bekesbourne data for a quadratic drift over neighbourhoods of (a) five terms (four lag intervals).

gram of log K at Broom's Barn Farm. The values of $\hat{\rho}(\mathbf{h})$ and $\hat{\gamma}(\mathbf{h})$ are listed in Table 5.1. In each graph there is one row of points at the top and one column of them on the right for the value 1.96 lying well away from all the others. This value might be regarded as an outlier, and it weakens the correlation. It also inflates the semivariance, especially so at the longer lag distances. This is evident in Figure 5.13 showing the computed correlograms and variograms with and then without this one point.

Screening

The importance of screening data as described in Chapter 2 using histograms and box-plots is now evident. All outliers must be regarded with suspicion and investigated. Erroneous values must be corrected or

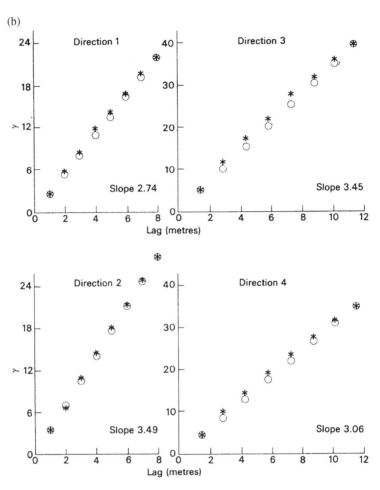

Figure 5.11 (*Cont.*) Variograms of a structural analysis of the Bekesbourne data for a quadratic drift over neighbourhoods of (b) nine terms (eight lag intervals).

Table 5.1 Autocorrelation coefficients and semivariances for log K with and without the largest value, 1.96.

Lag distance (m)	Autocorrelation		Semivariance	
	with 1.96	without 1.96	with 1.96	without 1.96
40	0.590	0.600	0.007 26	0.006 90
80	0.470	0.496	0.009 42	0.008 61
120	0.399	0.435	0.010 65	0.009 56
160	0.311	0.330	0.012 28	0.011 31

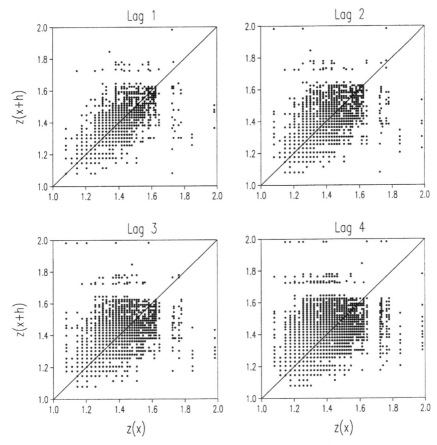

Figure 5.12 The h-scattergrams of $\log_{10} K$ at Broom's Barn Farm.

excised. Values that remain suspect are best removed, too. We recommend that where the values are correct, but recognizable as outliers, the data should be analysed with the outlier(s) both present and removed to judge the effect on the experimental semivariances. Some practitioners recommend removing them as a matter of course, arguing that they do not belong to the target population or to the realization of underlying process for the region. If by removing one or a very few values the skewness can be reduced then it is reasonable to do so to avoid transformation. For contaminated sites it is the extreme values that are often of interest. In this situation the variogram can be computed without the extreme values to ensure its stability, and then the values can be reinstated for estimation and other analyses. Some practitioners remove the 98th or even the 95th percentiles. This seems dangerous.

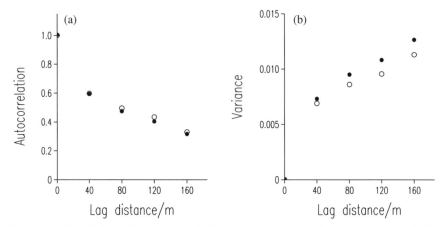

Figure 5.13 (a) Correlograms and (b) variograms from **h**-scattergrams of Figure 5.12 for lags of 1, 2, 3 and 4 sampling units: • complete data, ○ with extreme value removed.

Skew

We made the point in Chapter 2 that variances are unstable when data are strongly skewed (typically $g_1 \geqslant 1$). In these circumstances it is not only the extremes that cause instability. The solution we suggested in Chapter 2 applies equally to estimating semivariances. If the data are markedly skewed then transform them to approximate normality if you can and compute the variograms on the transformed data.

The estimates obtained with the usual computing formula for the variogram, equation (5.4), being variances, are sensitive to outliers and strong skewness in distributions. If the data are skewed then the confidence limits on the variogram are wider than they would otherwise be and as a result the semivariances are less reliable.

Even after transformation the tails of the distribution might be heavier than normal, and in some instances we might not want to transform where we are interested in a long upper tail, as in pollution studies. In this situation Cressie and Hawkins (1980) recommended a resistant estimator of the variogram. They discovered that the fourth root of the usual squared difference, namely,

$$y(\mathbf{h}) = [\{z(\mathbf{x}) - z(\mathbf{x} + \mathbf{h})\}^2]^{1/4}, \tag{5.27}$$

had a distribution close to normal with negligible skew. They therefore computed the mean, $\overline{y}(\mathbf{h})$, of all the individual sample values of $y(\mathbf{h})$. Taking equation (5.4) as a starting point, they computed

$$\overline{y}(\mathbf{h}) = \frac{1}{2m(\mathbf{h})} \sum_{i=1}^{m(\mathbf{h})} [\{z(\mathbf{x}_i) - z(\mathbf{x}_i + \mathbf{h})\}^2]^{1/4}. \tag{5.28}$$

The mean $\overline{y}(\mathbf{h})$ must be transformed back by

$$\hat{y}(\mathbf{h}) = \overline{y}(\mathbf{h}) \bigg/ 2\left\{0.457 + \frac{0.494}{m(\mathbf{h})} + \frac{0.045}{m^2(\mathbf{h})}\right\}. \tag{5.29}$$

McBratney and Webster (1986) experimented with this estimator, but they found it unsatisfactory for the kind of contaminated distributions that they envisaged in a pasture dotted with patches of dung and urine. Most recently, Lark (2000) has done a thorough study of robust estimators and compared that of Cressie and Hawkins (1980) with two others due to Dowd (1984) and Genton (1998), as well as with the standard estimator. He could find no general rule for choosing one rather than another or predicting their performance. The general approach, which we have stressed previously, is to examine the data, remove outliers if you believe them not to belong to the population that interests you, and to transform strongly skewed distributions to approximate normality before embarking on geostatistical analysis.

5.2.2 Sample size and design

In addition to the statistical distribution of the data, the reliability of the experimental variogram is also affected by the size of the sample (or its inverse, the density of data), and the configuration or design of the sample.

Hundreds, if not thousands, of experimental variograms are now displayed in published papers, reports, theses and books. They are derived from samples of as few as 24 individual measurements up to several thousands, though typically they are computed from 100–200 data. Those based on fewer than 50 data are often erratic sequences of experimental values with little or no evident structure. Figure 5.14 shows some examples. As the size of sample is increased such scatter decreases and the form of the variogram becomes clearer: the plotted points tend to be closer to an increasing line. Evidently the larger is the sample from which the variogram is computed, the more precisely is it estimated. In most instances, however, the precision is unknown, and we cannot determine how large a sample to take to achieve some desired precision. The classical formulae for determining confidence intervals cannot be applied unless the sampling itself is designed for the purpose, as suggested by Brus and de Gruijter (1994). Practitioners who attempt to assign error to their estimates based on these formulae are misguided. There are several reasons why:

(i) the same data are used more than once in each estimate;

(ii) the estimates are correlated;

(iii) the sampling is not sufficiently randomized.

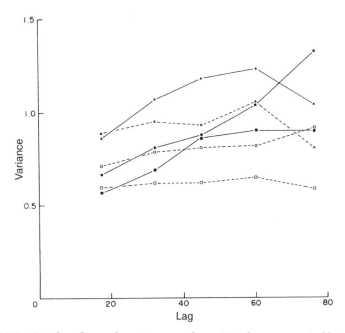

Figure 5.14 Graphs of sample variograms from 49 values generated by the same random process.

Before we proceed further we must be clear which variogram we are attempting to estimate from the experimental one. In Chapter 4 we identified two distinct functions, one the theoretical variogram and the other the local or regional variogram. The first is the variogram of the underlying stochastic process, whereas the local variogram is that of the particular realization in the region and called the non-ergodic variogram by Brus and de Gruijter (1994). An experimental variogram may contain error deriving from different realizations of the random function or from different samples of the particular realization, or both. In the first case the error arises from fluctuation in the generator, whereas in the second it arises from the sampling. We take the view here that for most practical purposes we are concerned with just one realization in the region, so we should try to estimate the sampling error expressed in the estimation variance or confidence limits.

Matheron (1965) gave a formula to provide a first approximation to the estimation variances of the local variogram:

$$\mathrm{var}[\hat{\gamma}_{R}(\mathbf{h})] \approx \frac{1}{N'}4\gamma(\mathbf{h})\sigma_{D}^{2}, \qquad (5.30)$$

where $\gamma(\mathbf{h})$ is the value of the theoretical variogram at lag $\mathbf{h}$, $\hat{\gamma}_{R}(\mathbf{h})$ is the estimate of the regional semivariance at that lag, and σ_{D}^{2} is the total variance in the region, i.e. the dispersion variance. Matheron describes N' as the number of points effectively used, i.e. the number of points that are superposed in the intersection of the region with itself when translated by the vector **h**. For a regular transect of length M it is the number of paired comparisons $(M - h)$ contributing to the estimate of $\gamma_R(\mathbf{h})$. It is from this that confusion has arisen about the number of observations needed to estimate the variogram reliably and, in particular, about the suggested minimum of 30–50 paired comparisons for any one $\hat{\gamma}_R(\mathbf{h})$ (Journel and Huijbregts, 1978). The advice seems to have been intended for one dimension, but unfortunately it has been applied widely in two dimensions and has given a false sense of security when computing variograms from small samples.

Muñoz-Pardo (1987) pursued Matheron's idea for estimating the estimation variances for variograms with a sill (bounded). He derived the following expression for the estimation variance of a semivariance:

$$\text{var}[\hat{\gamma}_R(\mathbf{h})] = \frac{1}{2S'^2} \int_S' \int_S f(\mathbf{x}, \mathbf{y}, \mathbf{h}) \, d\mathbf{x} d\mathbf{y}$$

$$+ \frac{1}{2N'^2(\mathbf{h})} \sum_{i=1}^{N'(\mathbf{h})} \sum_{j=1}^{N'(\mathbf{h})} f(\mathbf{x}_i, \mathbf{x}_j, \mathbf{h})$$

$$- \frac{1}{N'(\mathbf{h})S'} \sum_{i=1}^{N'(\mathbf{h})} \int_S' f(\mathbf{x}_i, \mathbf{x}, \mathbf{h}) \, d\mathbf{x}, \quad (5.31)$$

where

$$f(\mathbf{x}, \mathbf{y}, \mathbf{h}) = \{\gamma(\mathbf{x} - \mathbf{y} + \mathbf{h}) + \gamma(\mathbf{x} - \mathbf{y} - \mathbf{h}) - 2\gamma(\mathbf{x} - \mathbf{y})\}^2 \quad (5.32)$$

for any value of i and j. In equation (5.31) $\hat{\gamma}_R(\mathbf{h})$ denotes the estimated value of the regional variogram at lag **h**, S' is the area of intersection when the region is translated by the vector **h**, N' is the number of sampling points in the intersection, and **x** and **y** are two points that describe the region independently. Muñoz-Pardo solved the equation by numerical integration. He showed that the estimation variance depended on the effective range of the variogram in relation to the size of the region as well as on the size of the sample.

One way of obtaining confidence limits on variograms is by Monte Carlo methods (Webster and Oliver, 1992). There are two possible approaches, depending on which variogram (theoretical or regional) we are concerned with. If it is the first, then one simulates many realizations from a particular model and computes the experimental variogram of each, as did McBratney and Webster (1986), Taylor and Burrough (1986) and Shafer and Varljen (1990). The result will show the fluctuation arising from the generator, and the quantiles of the observed values for each lag would be reasonable estimates of the confidence intervals for new realizations.

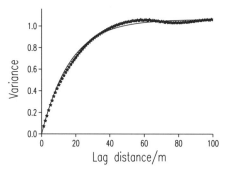

Figure 5.15 The exhaustive variogram computed from Figure 6.4(a).

Environmental scientists are more often concerned with single particular realizations, which they must sample, and so they are interested in
the sampling fluctuation. Here the Monte Carlo approach is to generate a
single large field of 'data' from a plausible model of the variation in the
region, sample repeatedly from it, and for each sample compute the sample variogram. The variation in the variograms thereby obtained will be
sampling fluctuation, and the quantiles of the semivariances may be used
as confidence limits on the regional variogram.

Webster and Oliver (1992) explored this approach by simulating large
autocorrelated random fields, which they then sampled on grids and transects of varying size and density with random starting points. We illustrate
the approach here with one of their examples.

A field of 65 536 random values on a 256×256 square grid with unit
interval was generated using sequential Gaussian simulation (Deutsch and
Journel, 1992) and an exponential variogram, see equation (6.26), with distance parameter $r = 16$ units. It is displayed in Figure 6.7(b). We can imagine it as one of exchangeable K in the soil. There are distinct patches with
large and small values showing that spatial dependence extends on average to about 50 units, which is about three times the distance parameter
(as explained in Chapter 6). The variogram from the exhaustive data was
close to the generating function (Figure 5.15). The field was then sampled
on regular square grids with the sample sizes and sampling intervals in
Table 5.2. No position was used more than once, and so no comparisons
were duplicated.

We computed the variograms for all the samples, and in Figure 5.16
we plot the results on the same set of axes. The variograms of the samples of 25 show the wide spread of estimates around the variogram of
the generating function. The dotted lines are the 90th percentiles, i.e. the
90% symmetrical confidence limits. The other graphs in the sequence (Figures 5.16), show how the confidence intervals narrow as the size of the
sample increases. A sample of 100 points appears to gives moderate confidence, but to attain satisfaction at least 144 measurements seem neces-

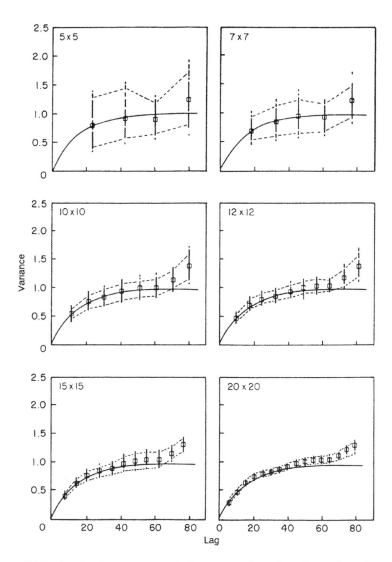

Figure 5.16 Semivariances computed on samples of various sizes from Figure 6.4(a) and 90% confidence limits obtained.

Table 5.2 Sample size, spacing and number of iterations.

Size	25	49	199	144	225	400
Interval	20	15	10	8	7	5
Iterations	100	100	100	64	49	25

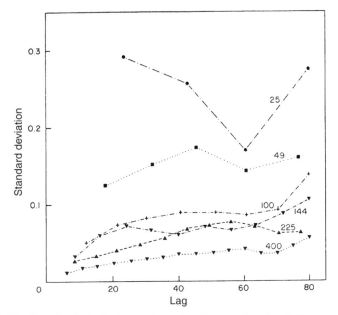

Figure 5.17   Standard deviations of semivariances for the different grid samplings from Figure 6.4(a).

sary. Increasing the sample to 225 points provides rather little improvement, while 400 data enable the variogram to be estimated with great precision.

The results can be summarized by plotting the standard deviation of the observed semivariances against the size of the sample (Figures 5.17). The standard deviation decreases, and the 90th percentiles narrow, with increasing sample size. We can also judge from Figure 5.17 approximately what size of sample to use to achieve some particular confidence.

The results show clearly that sample variograms from only 25 and 49 data have wide confidence intervals, and are therefore imprecise. Samples of 100 might be acceptable in some circumstances, and ones of 144 are likely to be adequate, at least with normally distributed isotropic data as in the generated fields. Variograms computed from samples of 225 will almost certainly be reliable, and samples of 400 seem extravagant. Based on this evidence, we recommend that you have no fewer than 100 sampling points and ideally 150 to estimate the variogram reliably in two dimensions if the variation is isotropic. For anisotropic variation we recommend at least 250 sample points because of the need to compute variograms in different directions.

The results also show the importance of interpreting N' of equation (5.30) correctly. Forty-nine points on a square grid gave us 84–240 paired comparisons, which would have seemed more than enough against the

30-50 comparisons regarded by some authorities as adequate. With 100 points there were 180-774 comparisons, and still the variograms were erratic.

Brus and de Gruijter (1994) viewed the problem differently. They pointed out that until you know the variogram accurately you cannot simulate realistic fields of values from which to sample. If the variogram used for the simulation has been estimated from few data then the realization generated might represent the true situation in the region poorly and lead to misleading confidence limits. They proposed a procedure based on classical sampling theory. For each lag, **h**, they repeatedly selected pairs of points, the first of each pair at random and the second determined by **h**. In the simplest design, simple random sampling (Chapter 2), the first point is chosen without regard to any other, and all points have equal chance of inclusion. If the variation is isotropic then the second point can be chosen by selecting it at distance $h = |\mathbf{h}|$ from the first but in a random direction. The mean of the individual squared differences obtained with equation (5.4), $\hat{y}(\mathbf{h})$, is an unbiased estimate of $y(\mathbf{h})$.

Since the pairs of points are chosen independently of one another the calculated squared differences are independent, and so the sampling variance of $\hat{y}(\mathbf{h})$ can be estimated by the classical formula. If we denote the squared difference at lag **h** by $d^2(\mathbf{h})$ then, following Cochran (1977), we can write the variance of the semivariance as

$$\text{var}[\hat{y}(\mathbf{h})] = \text{var}[0.5d^2(\mathbf{h})]/m(\mathbf{h})$$

$$= \frac{\frac{1}{4}\sum_{i=1}^{m(\mathbf{h})}\{d_i^2(\mathbf{h}) - \overline{d^2}(\mathbf{h})\}^2}{m(\mathbf{h})\{m(\mathbf{h}) - 1\}}, \tag{5.33}$$

where $\overline{d^2}(\mathbf{h})$ is the mean of the squared difference at lag **h**. Further, by choosing fresh pairs of points for each **h** or h, the estimates for the different lags are independent.

As we saw in Chapter 2, simple random sampling is inefficient, and the precision or efficiency can be improved by better design. Brus and de Gruijter (1994) elaborate the procedure for stratified sampling and give the formulae for the estimator and the estimation variance. The formulae can be modified for other designs.

The estimation variance has still to be converted into confidence limits, and for this one must assume a distribution. It is not immediately evident what that distribution should be. One might expect the individual $d^2(\mathbf{h})$ to be distributed as χ^2. Their means, however, are likely to approach normality with increasing $m(\mathbf{h})$ in accordance with the central limit theorem. Brus and de Gruijter calculated limits on this assumption but found that it was not entirely satisfactory for the fairly small $m(\mathbf{h})$ in their study: they obtained several negative lower limits at the 90% level, suggesting that the confidence interval is not symmetric, at least for the small samples they took. This contrasts with our finding, with larger samples, that limits were approximately symmetrical.

Despite this weakness, the method proposed by Brus and de Gruijter gives sound unbiased estimates of the sampling variance of $\gamma(\mathbf{h})$, but large samples are needed to obtain precise estimates. In addition, the sampling scheme with pairs of points scattered irregularly and unevenly is inefficient for subsequent kriging (Chapter 8).

Although the above approaches to the problem are different, both show that the confidence intervals are very wide with small samples: you need as large a sample as possible to estimate the variogram reliably.

5.2.3 Sample spacing

In general, as the size of sample increases so the spacing between sampling points decreases for a given region. Nevertheless, we cannot simply allow the sample spacing to be dictated by the size of sample. The spacing must relate to the scale or scales of variation in the region. Otherwise we might sample too sparsely to identify correlation. We should therefore know roughly the spatial scale of variation in Z so as to choose a sensible sampling density.

Some variables, such as vegetation, have visible patterns, and their spatial scales are obvious. Many properties of soil, rocks, atmosphere and water, on the other hand, are invisible, and so one cannot judge the spatial scales on which they vary without first sampling. They can also vary on scales that differ by several orders of magnitude simultaneously, as described in Chapter 4. In some instances an approximate scale of variation can be judged from that of other features, such as landform or vegetation, but often it is more elusive.

Let us consider the following situations.

1. *Terra incognita.* If we know nothing of the pattern or scale of the variation then it is difficult to choose a sampling interval rationally. A large interval might be too large to capture the autocorrelation. If we choose a small interval then we might have to restrict the area sampled to stay within a budget and fail to estimate long-range variation. If we were to sample a whole region densely and the variation turned out to be entirely long-range then we should have wasted money trying to estimate short-range variation. We want some means of estimating, even roughly, the spatial scale of variation effectively and economically.

2. We have data from a previous survey, but their experimental variogram(s) seem(s) flat, or *pure nugget* as we shall call them in Chapter 6, i.e. there is no evident spatial correlation. If the variables are

continuous then we can assume that the correlation range is less than the smallest sampling interval. We can know no more than that.

3. We have variograms with apparent 'structure', but feel that some parts of the region are undersampled and others oversampled, and that some sampling points could be positioned more effectively to optimize estimation.

The problems faced in 1 and 2 can be resolved by starting with a nested survey and analysis, which we now describe.

5.3 THEORY OF NESTED SAMPLING AND ANALYSIS

The model of nested variation is based on the notion that a population can be divided into classes at two or more categoric levels in a hierarchy. The population can then be sampled using a multi-stage or nested scheme so as to estimate the variance at each level. The population is divided initially into classes at stage 1, and these are subdivided at stage 2 into subclasses, which in turn can be subdivided further at stage 3 to give finer classes, and so on, to form a nested or hierarchical classification with m stages. In each case the class at the lower level is contained completely within the one immediately above it, and each sampling point is contained in one and only one class at each and every level. The system is a strict hierarchy, and a single observation embodies variation contributed by each of the stages, including an unresolved variance within the classes at the finest level of resolution. We can estimate these contributions to the variance by a hierarchical analysis of variance.

Youden and Mehlich (1937) saw that for an attribute distributed in space the stages could be represented by a hierarchy corresponding to different distances. They adapted classical multi-stage sampling so that each stage in the hierarchy represented a distance between sampling points. They sampled at random, with only the distances between pairs fixed, and so the random effects model, model II of Marcuse (1949), is appropriate for the analysis of variance.

For a design with m stages the model of the variation is

$$Z_{ijk...m} = \mu + A_i + B_{ij} + \cdots + \varepsilon_{ijk...m}, \qquad (5.34)$$

where $z_{ijk...m}$ is the value of the mth unit in ..., in the kth class at stage 3, in the jth class at stage 2, and in the ith class at stage 1. The general mean is μ; A_i is the difference between μ and the mean of class i in the first category; B_{ij} is the difference between the mean of the jth subclass in class i and the mean of class i; and so on. The final quantity $\varepsilon_{ijk...m}$ represents the deviation of the observed value from its class mean at the last stage of subdivision. The quantities $A_i, B_{ij}, C_{ijk}, \ldots, \varepsilon_{ijk\ldots m}$ are assumed to be independent random variables associated with stages $1, 2, 3, \ldots, m$, respectively, having means of zero and variances $\sigma_1^2, \sigma_2^2, \sigma_3^2, \ldots, \sigma_m^2$. The latter are the components of variance for the respective stages. They are estimated according to the scheme in Table 5.3. The quantities $n_1, n_2, n_3, \ldots, n_m$, in the table are the numbers of subdivisions of each class at the several levels. If for each stage, say j, n_j is constant then the design is balanced. All the $n_j, j = 1, 2, \ldots, n_m$, are known for any particular design, and so we can determine the components of variance for all stages in the classification and the residual variance from the right-hand column of Table 5.3.

The individual component for a given stage measures the variation attributable to that stage, and together they sum to the total variance:

$$\sigma^2 = \sigma_1^2 + \sigma_2^2 + \sigma_3^2 + \cdots + \sigma_m^2. \tag{5.35}$$

The components of variance for each spacing from this analysis reveal over what part of the spatial scale most of the variation occurs. The particular merit of the method is that a wide range of spatial scales can be covered in a single analysis.

5.3.1 Link with regionalized variable theory

Although the hierarchical analysis of variance derives from classical statistics, Miesch (1975) has shown its links with geostatistics. He also showed that it can provide a first approximation to the variogram if the components of variance are accumulated, starting with the smallest spacing. For the m stages of subdivision we have the corresponding distances $d_1, d_2, \ldots, d_m$, where d_1 is the shortest distance at the mth stage and d_m the largest distance at the first stage. Then the equivalence is given by

$$\sigma_m^2 = \gamma(d_1),$$
$$\sigma_{m-1}^2 + \sigma_m^2 = \gamma(d_2), \tag{5.36}$$
$$\sigma_{m-2}^2 + \sigma_{m-1}^2 + \sigma_m^2 = \gamma(d_3),$$

and so on. In practice, the components tend to be fairly rough estimates of the true semivariances because each is usually based on few degrees of freedom. They can also be biased because they are not entirely independent.

The values $\gamma(d_i)$ are the equivalent semivariances. When plotted against distance they provide a first approximation to the variogram. The result might be rough, but it indicates how Z varies in space in the region over several orders of magnitude of distance in a single analysis and with modest sampling. For this reason it is ideal for reconnaissance where little or

Table 5.3 Hierarchical analysis of variance: parameters estimated.

Source	Degrees of freedom	Parameters estimated
Stage 1	$n_1 - 1$	$\sigma_m^2 + n_m \sigma_{m-1}^2 + \cdots + n_m n_{m-1} \ldots n_3 \sigma_2^2 + n_m n_{m-1} \ldots n_3 n_2 \sigma_1^2$
Stage 2	$n_1(n_2 - 1)$	$\sigma_m^2 + n_m \sigma_{m-1}^2 + \cdots + n_m n_{m-1} \ldots n_3 \sigma_2^2$
Stage 3	$n_1 n_2(n_3 - 1)$	$\sigma_m^2 + n_m \sigma_{m-1}^2 + \cdots + n_m n_{m-1} \ldots n_4 \sigma_3^2$
$\cdots$	$\cdots$	$\cdots$
Stage $m - 1$	$n_1 n_2 n_3 \ldots (n_{m-1} - 1)$	$\sigma_m^2 + n_m \sigma_{m-1}^2$
Stage m (residual)	$n_1 n_2 n_3 \ldots n_{m-1}(n_m - 1)$	σ_m^2
Total	$n_1 n_2 n_3 \ldots n_{m-1} n_m - 1$	

Table 5.4   Components of variance of pH in two soil series in Broome County.

Stage	Spacing (m)	Degrees of freedom	Culvers series 0-15 cm		Sassafras series 0-15 cm	
			Estimated component	Percentage of variance	Estimated component	Percentage of variance
1	1600.0	8	0.028 19	39.7	0	0
2	305.0	9	0.023 40	32.9	0.044 40	60.3
3	30.5	18	0.005 52	7.8	0.006 98	9.8
4	3.5	36	0.013 91	19.6	0.022 25	30.2

nothing is known. Once the spatial scale is known then subsequent survey can be planned to estimate the variogram precisely (Oliver and Webster, 1986) or to plan a more general survey over a larger area. Alternatively, the analysis might show that all the variation occurs over very short distances, and that attempting to measure spatial correlation and map the variable(s) is pointless or would be too costly.

5.3.2   Case study: Youden and Mehlich's survey

We illustrate the technique with an example from Youden and Mehlich's (1937) original paper. The authors' sampling scheme to survey the soil in Broome County in New York State had four stages (Table 5.4). They applied it to two soil series: the Culvers and the Sassafras. On each soil type they selected nine primary centres 1.6 km apart forming level 1 in the hierarchy. At the next level (level 2) one subcentre was selected 305 m away from each of the main centres (18 locations). At level 3 two sampling points were chosen 30.5 m from the main centre and the subcentre (36 locations). At level 4 each site was replicated with another 3.05 m away, to give 72 sampling points in all. The survey was fully balanced, so that all classes at a particular level were subdivided equally to form the hierarchy (Figure 5.18). The progression of the sampling intervals was geometric, and as a result the authors felt able to regard the components of variance as independent, thereby allowing confidence intervals to be determined. At each sampling point the pH was determined on soil taken from a depth of 0–15 cm.

For each soil series the variation associated with each sampling interval was determined by a nested analysis of variance as in Table 5.3. First the sums of squares of the deviations of the means of the classes at level 1 from the general mean were computed and then each was multiplied by the number of observations that make up the class mean. For each class at level 2, the difference between its mean and the mean of the class to which

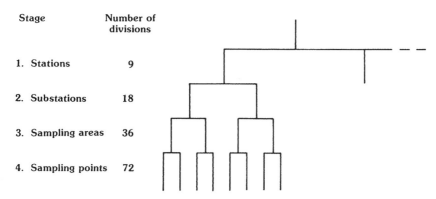

Stage	Number of divisions
1. Stations	9
2. Substations	18
3. Sampling areas	36
4. Sampling points	72

Figure 5.18 Topological structure of a balanced nested sampling design.

it belongs in level 1 was squared and multiplied by the number of observations in that class. The sum of these values is the appropriate sum of squares. This was repeated for each stage, and the sums of squares of the individual levels sum to the total sum of squares. The mean squares were obtained by dividing the sums of squares of each stage by the appropriate degrees of freedom (Table 5.3). The mean square at each level, apart from the lowest, contains a unique contribution to the variance from that level, plus contributions from the components at all levels below (Table 5.3). For instance, the unique contribution to the variance at level 2 (Table 5.3) is $n_m n_{m-1} \ldots n_3 \sigma_2^2$. This enables each component to be determined separately from its mean square. For a balanced design the values of each component can be tested to judge whether it is larger than zero by computing the F ratio:

$$F = \frac{\text{mean square at level } m}{\text{mean square at level } m + 1}. \tag{5.37}$$

Table 5.4 gives the results of the analysis. The components of variance can now be accumulated, starting with that for the lowest level, and plotted against sample spacing (Figure 5.19). This gives a first approximation to the variogram, a reconnaissance variogram. Figure 5.19 shows the accumulated components of variance for the Culvers and Sassafras series plotted against distance on a logarithmic scale. The variance for the Culvers series increases substantially as the distance between sampling points increases and without limit. The sample variance for the Sassafras series does reach a maximum, and we might therefore treat the variation as second-order stationary. If we project the variances to spacings less than the smallest, 3 m, then both seem to approach limits larger than 0. Later, in Chapter 6, we shall recognize this as *nugget variance*, and we shall explain it there.

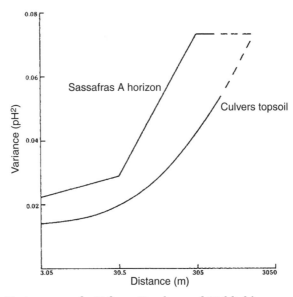

Figure 5.19 Variograms of pH from Youden and Mehlich's survey for Culvers soil and Sassafras soil by accumulating the components of variance. Note the logarithmic scale for distance on the abscissa.

Table 5.5 Nested sampling design for determining the scale of spatial variation in the soil of the Wyre Forest.

Stage	Sampling interval (m)	Number of sampling points
1	600	9
2	190	18
3	60	36
4	19	72
5	6	108

5.3.3 Unequal sampling

The sampling designs described above are fully balanced in the sense that all classes at each particular stage are subdivided equally. For the particular design in Broome County the sample size doubles with each additional stage in the hierarchy after the first. To achieve good spatial resolution may require many stages and result in prohibitively expensive sampling for reconnaissance. It would to some extent defeat the object whereby one is trying to obtain preliminary information for modest effort. As it happens, full replication at each stage is unnecessary because the mean

Table 5.6   Hierarchical analysis of variance: unbalanced design.

Source	Degrees of freedom	Parameters estimated
Level 1	$f_1 - 1$	$u_{1,1}\sigma_1^2 + u_{1,2}\sigma_2^2 + u_{1,3}\sigma_3^2 + \sigma_4^2$
Level 2	$f_2 - f_1$	$u_{2,2}\sigma_2^2 + u_{2,3}\sigma_3^2 + \sigma_4^2$
Level 3	$f_3 - f_2$	$u_{3,3}\sigma_3^2 + \sigma_4^2$
Residual	$N - f_3$	σ_4^2
Total	$N - 1$	

squares for the lower stages are estimated more precisely than those for the higher ones. Economy can be achieved by replicating only a proportion of the sampling centres in the lower stages. Oliver and Webster (1986) used five stages, but in more recent applications, Webster and Boag (1992), Badr *et al.* (1993) and Oliver and Badr (1995) have used seven. The resulting schemes are unbalanced, and this makes estimating the components somewhat more complex because the coefficients of the components are no longer simple multiples of the number of divisions in the levels as they are in Table 5.3, which must be replaced by a table such as Table 5.6 for a sample of size N.

Gower (1962) provided formulae for calculating the coefficients, u_{ij}, and they were included in the sixth edition of Snedecor and Cochran's (1967) standard text (but not in the later editions). They can all be expressed in the following single formula:

$$u_{ij} = \frac{1}{d_i}\left\{\sum_{k=1}^{C_i}\sum_{p=1}^{c^i_{jk}}\frac{n_{pj}^2}{n_{ik}} - \sum_{k=1}^{C_{i-1}}\sum_{p=1}^{c^{i-1}_{jk}}\frac{n_{pj}^2}{n_{i-1,k}}\right\}, \tag{5.38}$$

where u_{ij} is the coefficient at level i for the jth component; C_i is the number of groups at the ith level; c^i_{jk} is the number of subgroups in level j within the kth group at level i; n_{pj} is the number of individual sampling points in the pth subgroup in level j (within group k at level i), with $i \leqslant j$; d_i is the number of degrees of freedom at level i.

One consequence of the lack of balance is that the coefficients for a given component in the expected values for the mean squares are in general different in different levels. As a result one cannot use a simple F ratio to test whether a component, σ_j^2, is significantly greater than 0.

5.3.4   Case study: Wyre Forest survey

A survey of the soil in the Wyre Forest, in the English Midlands, illustrates the unbalanced nested design (Oliver and Webster, 1987). The variograms from an earlier survey were flat; all the variation in the properties examined occurs within 167 m, the average distance between sampling sites in

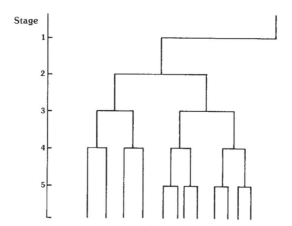

Figure 5.20   Topology for one main centre of the unbalanced nested sampling as implemented in the soil survey of the Wyre Forest (Oliver and Webster, 1987).

that survey. The nested survey was designed to discover how the variation is distributed over distances less than 167 m. The scheme had five stages covering a range of sampling intervals from 6 m increasing in a geometric progression of approximately threefold increments (Table 5.5) to 600 m. This design was expected to encompass most of the spatial variation, and to ensure that the components of variance could be regarded as independent. The 600 m interval was incorporated in case there were long-range spatial structures.

Nine main centres were located at the nodes of a 600 m square grid oriented randomly over the region. All other points were then placed on random orientations from these as follows to comply with the random effects model. From each grid node a second site was chosen 190 m away to provide the second stage. From each of the now 18 sites another point was chosen 60 m away (stage 3). The procedure was repeated at stage 4 to locate points 19 m away from those of stage 3, giving 72 points. At the fifth stage just half of the fourth-stage points were replicated by sampling 6 m away. This gave a sample of 108 points rather than 144 for a fully balanced survey. Table 5.5 summarizes the design, Figure 5.20 illustrates the hierarchical structure used for one centre, and Figure 5.21 shows the configuration of sampling points for one first-stage centre. The design achieved a 25% economy in effort compared with a fully balanced scheme. Figure 5.22 shows the economy possible with even more stages. At each sampling point several properties of the soil were recorded at four fixed depths in the soil profile: 0–5 cm (1), 10–15 cm (2), 25–30 cm (3), and 50–55 cm (4).

Each variable was analysed according to the scheme outlined in Table 5.6. The estimated components of variance for sand content at the four

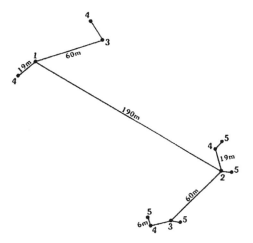

Figure 5.21   Sampling plan for one main centre of the Wyre Forest survey.

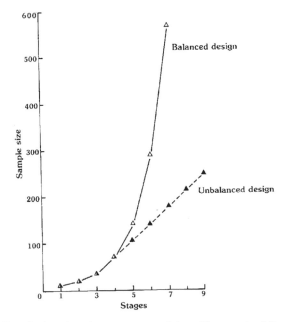

Figure 5.22   Graph showing the economy achieved by not doubling the sampling at every level in the hierarchy.

depths are listed in Table 5.7. Figure 5.23 shows the accumulated components of variance for each depth in the profile plotted against separating distance on a logarithmic scale to give first approximations to the variograms. The graph shows that at least 80% of the variation at the

Table 5.7 Components of variance and percentage variance of sand content of the soil at four depths contributed by each stage in the survey of the Wyre Forest.

	Component of variance Depth (cm)				Percentage of variance Depth (cm)			
Stage	0–5	10–15	25–30	50–55	0–5	10–15	25–30	50–55
1	37.45	21.26	34.17	40.40	12.0	6.5	7.6	7.5
2	−47.25	−60.49	−80.94	−104.7	0	0	0	0
3	85.79	139.90	171.20	317.10	27.5	43.0	37.9	51.5
4	147.40	113.80	155.40	0.70	47.2	35.0	3.4	0.1
5	41.69	50.21	90.46	251.70	13.3	15.4	20.0	40.9

four depths for sand content occurs within 60 m; this was the case for all of the other properties. Stages 1 and 2, i.e. distances of 190–600 m and 60–190 m, respectively, account for less than 20% of the variation. The estimated components of variance for stage 2 were generally negative. This suggests that either there is some repetition in soil character at that distance, or that the components estimate zero because there is no contribution to the variance at this stage. The confidence limits of the components are wide, and so we cannot be sure how to interpret these negative values. Even at stage 5 there is still a considerable contribution to the total variance. This represents the unresolved variation within 6 m plus errors of measurement.

As described above, the experimental variogram depends on the spatial scale over which we measure it. If a large extent is covered with wide sampling intervals then all of the variance might appear as nugget. Alternatively, if small intervals are chosen to resolve the short-range variance then the sampling required to estimate the contributions to the larger distances would be too costly. A nested survey identifies the scale at which most of the variation occurs at the level of our investigation. The reconnaissance variograms for the soil properties of the Wyre Forest showed that most of the spatial variation occurred over distances less than 60 m.

From this information we could design a survey to estimate the variogram more precisely by linear sampling. We did so using ten transects each 100 m long and one of 500 m with a sampling interval of 5 m. The conventional variograms that resulted showed correlation extending to little more than about 40 m.

We could have used the results of the nested survey to design an overall survey with a maximum sampling interval equal to half the correlation range identified. This would have been 30 m. In the event, having estimated the variograms more precisely along transects and established an

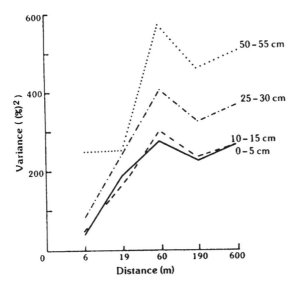

Figure 5.23 Variograms of soil properties in the Wyre Forest obtained by accumulating the components of variance, with the lag distance put on a logarithmic scale.

effective range of 40 m, we sampled at a 20 m interval from which to interpolate for mapping.

5.3.5 Summary

Nested survey and analysis can reveal the spatial scale(s) of variation in a region. The estimates of the components may be somewhat biased, and with the modest sampling that we illustrate above the estimates will have fairly large errors. As a first step in describing variation, however, these are of secondary importance. Armed with the information they provide, one can plan a second stage of survey to estimate the variogram precisely over the range that matters. The results from nested survey could be used to plan a regional survey if one particular component proved dominant.

6

Modelling the variogram

In Chapter 5 we saw that when we compute an empirical variogram we obtain an ordered set of values, the experimental or sample variogram, consisting of $\hat{y}(\mathbf{h}_1), \hat{y}(\mathbf{h}_2), \ldots$, at particular lags, $\mathbf{h}_1, \mathbf{h}_2, \ldots$. This variogram summarizes the spatial relations in the data. We usually want more than that, however; we want a variogram to describe the variance of the region. Each calculated semivariance for a particular lag is only an estimate of a mean semivariance for that lag. As such it is subject to error. This error, arising largely from sampling fluctuation, gives the experimental variogram a more or less erratic appearance (see Figure 5.16). The true variogram representing the regional variation is continuous, and it is this variogram that we should really like to know. We can use our observed values as approximations to the function by imagining a curve passing through them—or in two dimensions a surface, for the variogram of a two-dimensional field is itself two-dimensional. How closely should we attempt to follow the experimental variogram? Answering this is difficult because we do not know how much of the observed fluctuation is due to error and how much is structural.

The solution usually taken is that of Occam's razor; namely, fit the simplest function that makes sense, subject to certain mathematical constraints which are considered below. We ignore the point-to-point fluctuation and concentrate on the general trends.

Another reason for fitting a continuous function is to describe the spatial variation so that we can estimate or predict values at unsampled places and in larger blocks of land optimally by kriging (see Chapter 8). This requires semivariances at lags for which we have no direct comparisons, and we must be able to calculate these from such a function. The function must therefore be mathematically defined for all real $\mathbf{h}$.

There are a few principal features that a function must be able to represent. These include:

(1) a monotonic increase with increasing lag distance from the ordinate of appropriate shape;

(2) a constant maximum or asymptote, or 'sill';

(3) a positive intercept on the ordinate, or 'nugget';

(4) periodic fluctuation, or a 'hole';

(5) anisotropy.

6.1   LIMITATIONS ON VARIOGRAM FUNCTIONS

6.1.1   Mathematical constraints

Not any close-fitting function will serve. The model we choose must describe random variation, and the function must be such that it will not give rise to 'negative variances' of combinations of random variables. This is explained below.

Let $z(\mathbf{x}_i)$, $i = 1, 2, \ldots, n$, be a realization of the random variable $Z(\mathbf{x})$ with covariance function $C(\mathbf{h})$ and variogram $\gamma(\mathbf{h})$. Now consider the linear sum

$$y = \sum_{i=1}^{n} \lambda_i z(\mathbf{x}_i),$$

where the λ_i are any arbitrary weights. The variable Y from which y derives is itself random with variance

$$\text{var}[Y] = \sum_{i=1}^{n} \sum_{j=1}^{n} \lambda_i \lambda_j C(\mathbf{x}_i - \mathbf{x}_j), \tag{6.1}$$

where $C(\mathbf{x}_i - \mathbf{x}_j)$ is the covariance of Z between $\mathbf{x}_i$ and $\mathbf{x}_j$. The variance of Y may be positive or zero; but it may not be negative. The right-hand side of equation (6.1) must ensure this. The covariance function, $C(\mathbf{h})$, must be *positive semidefinite*. Equation (6.1) can be written

$$\text{var}[Y] = \boldsymbol{\lambda}^{\mathsf{T}} \mathbf{C} \boldsymbol{\lambda} \geqslant 0, \tag{6.2}$$

where $\boldsymbol{\lambda}$ is the vector of weights and $\mathbf{C}$ is the matrix of covariances. If the latter is positive semidefinite then so is the covariance function. In fact, since we are dealing with 'variables', the variance cannot be zero, and so $C(\mathbf{h})$ must be positive definite.

If the covariance does not exist, because the variable is intrinsic only and not second-order stationary, then we rewrite equation (6.1) as

$$\text{var}[Y] = C(\mathbf{0}) \sum_{i=1}^{n} \lambda_i \sum_{j=1}^{n} \lambda_j - \sum_{i=1}^{n} \sum_{j=1}^{n} \lambda_i \lambda_j \gamma(\mathbf{x}_i - \mathbf{x}_j), \tag{6.3}$$

where $\gamma(\mathbf{x}_i - \mathbf{x}_j)$ is the semivariance of Z between $\mathbf{x}_i$ and $\mathbf{x}_j$. The first term on the right-hand side of equation (6.3) contains $C(\mathbf{0})$, the covariance at lag **0**, which we do not know, but we can eliminate it by making the weights sum to 0 without loss of generality. Then

$$\text{var}[Y] = -\sum_{i=1}^{n}\sum_{j=1}^{n}\lambda_i\lambda_j\gamma(\mathbf{x}_i - \mathbf{x}_j).\qquad(6.4)$$

This may not be negative either; but notice the minus sign. So, the variogram must be *conditional negative semidefinite* (CNSD), the condition being that the weights in equation (6.4) sum to zero.

Only functions that ensure non-zero variances may be used for variograms. They are called *authorized* models or functions in much of the literature.

6.1.2 Behaviour near the origin

The way in which the variogram approaches the origin is determined by the continuity (or lack of continuity) of the variable, $Z(\mathbf{x})$, itself. We may distinguish the following features, which are illustrated in Figure 6.1. As mentioned in Chapter 4, the variogram is symmetric about the origin, and for reasons that will become evident both halves appear in this figure.

Positive intercept. The semivariance at $|\mathbf{h}| = 0$ is by definition 0. It often happens, however, that any line or surface projected through the experimental values to the ordinate intersects it at some positive value, as in Figure 6.1(a). This implies a discontinuity in $Z(\mathbf{x})$. The feature appeared often in gold mining, and the mining engineers attributed it to the spatially independent occurrence of gold nuggets in ore bodies. They called the phenomenon the 'nugget effect' and the intercept 'nugget variance'. It is easy to imagine discontinuities arising from the dispersal of small

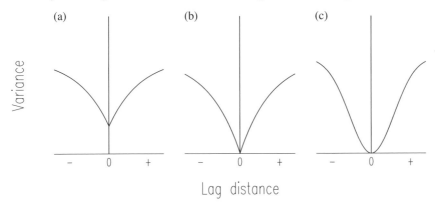

Figure 6.1 Behaviour of the variogram near the origin: (a) positive intercept (nugget); (b) linear approach, not differentiable; (c) continuous and differentiable.

nuggets of gold in a large body of rock. The same might be true for certain features of the soil, such as stones and concretions among the fine earth. Discontinuities in the soil's physical and chemical properties are harder to imagine, and any apparent nugget variance usually arises from errors of measurement and spatial variation within the shortest sampling interval.

Linear approach. The variogram may approach the origin approximately linearly with decreasing lag distance:

$$\gamma(\mathbf{h}) \approx b|\mathbf{h}| \qquad \text{as } |\mathbf{h}| \rightarrow 0, \qquad (6.5)$$

where b is the gradient. The variogram passes through the origin, as in Figure 6.1(b) (unlike in Figure 6.1(a)), but it is discontinuous there: its gradient changes abruptly from negative to positive. Nevertheless, it signifies continuity in $Z(\mathbf{x})$ itself, and because

$$\lim E[\{Z(\mathbf{x}) - Z(\mathbf{x} + \mathbf{h})\}^2] = 0 \qquad \text{as } |\mathbf{h}| \rightarrow 0, \qquad (6.6)$$

$Z(\mathbf{x})$ is often said to be 'mean-square' continuous. It is not differentiable, however, nor is the process it describes, because it is random (see Chapter 4).

Parabolic approach. Figure 6.1(c) illustrates the situation in which a variogram is parabolic at the origin; it passes smoothly through the origin with a gradient of 0 there, so that

$$\gamma(\mathbf{h}) = b|\mathbf{h}|^2 \qquad \text{as } |\mathbf{h}| \rightarrow 0. \qquad (6.7)$$

The variogram is twice differentiable at the origin, and $Z(\mathbf{x})$ is itself differentiable: it varies smoothly, and it is no longer random. The exponent 2 represents a strict limit to power functions for describing random processes.

6.1.3   Behaviour towards infinity

The way that a variogram behaves with increasing lag distance is constrained by

$$\lim \frac{\gamma(\mathbf{h})}{|\mathbf{h}|^2} = 0 \qquad \text{as } |\mathbf{h}| \rightarrow \infty. \qquad (6.8)$$

The variogram must increase less than the square of the lag distance as the latter approaches infinity; if it does not then the process is not entirely random. The limit is shown in Figure 6.2, in which the parameter α in the power function is set to 2. Any function that increases more is not CNSD and so is not compatible with the intrinsic hypothesis.

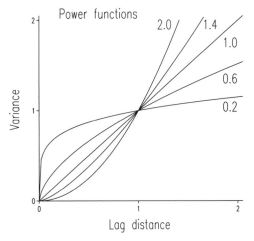

Figure 6.2 Graphs of the power function, $\gamma(h) = wh^\alpha$, with $\alpha = 0.2, 0.6, 1.0,$ 1.4 and 2.0 (the limiting value for α), and with w set to 1.

6.1.4 Drift and trend

A raw variogram that appears parabolic at the origin suggests that there is local *drift*, i.e. short-range deterministic variation. A variogram that increases faster than $|\mathbf{h}|^2$ suggests that there is long-range trend, again deterministic in the statistical sense. These features are described in Chapter 4. In both cases the expectation of $Z(\mathbf{x})$ is not stationary but depends on position $\mathbf{x}$:

$$\mathrm{E}[Z(\mathbf{x})] = m(\mathbf{x}). \tag{6.9}$$

6.2 AUTHORIZED MODELS

There are two main families of simple functions that encompass the features listed above and that are CNSD. One represents unbounded variation, the other bounded. We deal with them in turn in their isotropic form, so that the lag vector $\mathbf{h}$ becomes a scalar measure in distance only, h. All of the ones that we describe are used in practice.

6.2.1 Unbounded random variation

The idea of unbounded, i.e. infinite, variance may seem strange. After all, we live on a finite earth, and there must be some limit to the amount of variation in the soil. Yet the evidence from surveys of small parts of the planet suggests that if we were to increase the region surveyed we should encounter ever more variation; our extrapolation of the experimental variogram is one that continues to increase.

The simplest models for unbounded variation are the power functions:

$$\gamma(h) = wh^{\alpha} \quad \text{for } 0 < \alpha < 2, \tag{6.10}$$

where w describes the intensity of variation, and α describes the curvature. If $\alpha = 1$ then the variogram is linear, and w is simply the gradient. If $\alpha < 1$ then the variogram is convex upwards. If $\alpha > 1$ then the variogram is concave upwards. The limits 0 and 2 are excluded. If $\alpha = 0$ then we are left with a constant variance for all $h > 0$; if $\alpha = 2$ then the function is parabolic with gradient 0 at the origin and represents differentiable variation in the underlying process, which is not random, as mentioned above.

Figure 6.2 shows examples with several values of α, including the upper bound, $\alpha = 2$; at the lower limit, $\alpha = 0$ would represent white noise, and hence discontinuous variation. Nevertheless, some experimental variograms seem flat, and we return to this matter below.

One way of looking at these unbounded functions is to consider Brownian motion in one dimension. Suppose a particle moves in this dimension with a velocity or momentum at position $\mathbf{x} + \mathbf{h}$ that depends on its velocity or momentum at a close previous position $\mathbf{x}$. It can be represented by the equation

$$Z(\mathbf{x} + \mathbf{h}) = \beta Z(\mathbf{x}) + \varepsilon, \tag{6.11}$$

where ε is an independent Gaussian random deviate and β is a parameter. At its simplest $\beta = 1$, and its variogram is then

$$2\gamma(\mathbf{h}) = E[\{Z(\mathbf{x} + \mathbf{h}) - Z(\mathbf{x})\}^2] = |\mathbf{h}|^k. \tag{6.12}$$

If the exponent k in equation (6.12) is 1 then we obtain the linear model, with $\gamma(|\mathbf{h}|) \to \infty$ as $|\mathbf{h}| \to \infty$. This is also known as a *random walk* model.

In ordinary Brownian motion the εs are independent of one another. However, if the εs in equation (6.11) are spatially correlated then a trace is generated that is smoother than that of pure Brownian motion. The exponent, k, now exceeds 1, and the curve is concave upwards. If, on the other hand, the εs are negatively correlated then a trace is generated that is rougher, or 'noisier', than that of pure Brownian motion. The exponent k in equation (6.12) is now less than 1, and the curve is convex upwards.

If the εs are perfectly correlated then $k = 2$ and the trace is completely smooth, i.e. there is no longer any randomness. As $k \to 0$, the noise increases until in the limit we have white noise, or pure nugget, as described in Chapter 4.

Priestley (1981) gives a much more comprehensive account of these random processes. Chapter 3 is especially relevant, and we must leave the reader to pursue the matter there.

6.2.2 Bounded Models

In our experience bounded variation is more common than unbounded variation, and the variograms have more varied shapes. In most of these models the variance has a maximum, which is the *a priori* variance of the process, known in geostatistics as the *sill* variance. The variogram may reach its sill at a finite lag distance, the *range*. Alternatively, the variogram may approach its sill asymptotically. In some models the semivariance reaches a maximum, only to decrease again and perhaps fluctuate about its *a priori* variance. These variograms represent second-order stationary processes and so have equivalent covariance functions. They are illustrated in Figures 6.3 and 6.4.

Bounded linear

The simplest function for describing bounded variation consists of two straight lines, see Figure 6.3(a). The first increases and the other has a

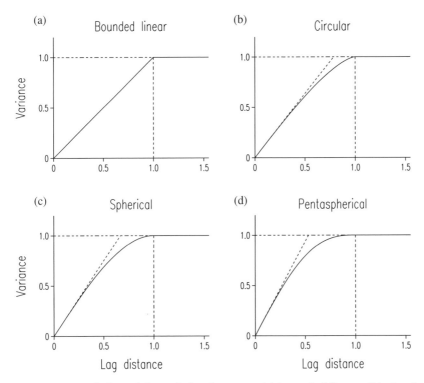

Figure 6.3 Bounded models with fixed ranges: (a) bounded linear; (b) circular; (c) spherical; (d) pentaspherical.

constant variance:

$$y(h) = \begin{cases} c\left(\dfrac{h}{a}\right) & \text{for } h \leqslant a, \\ c & \text{for } h > a, \end{cases} \qquad (6.13)$$

where c is the sill variance and a is the range. Evidently its slope at the origin is c/a. It is CNSD in one dimension ($\mathbb{R}^1$) only; it may not be used to describe variation in two and three dimensions.

We can derive the variogram for the bounded linear model heuristically as follows. We start with a stationary 'white noise' process, $Y(x)$, in one dimension, i.e. a random process with random variables at all positions along a line but in which there is no spatial dependence or autocorrelation. It has a mean μ and variance σ_Y^2. Suppose that we pass the process through a simple linear filter of finite length a to obtain

$$Z(x) - \mu = \frac{1}{a} \int_x^{x+a} Y(v)\, \mathrm{d}v. \qquad (6.14)$$

Thus, we average $Y(x)$ within the interval a to obtain the corresponding $Z(x)$. Consider now the variable $Z(x)$ derived from two segments of length a of the process $Y(x)$, one from x_1 to x_2 and the other from x_3 to x_4. They may overlap or not, as below.

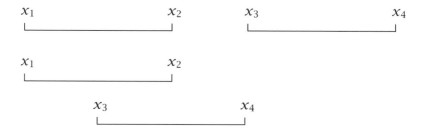

Evidently, if the two segments do not overlap, as in the upper example, then we should expect their means in $Z(x)$ to be independent. But if they do overlap, as in the lower example, then they will share some of the original white noise series; their means will not be independent, and we should expect some autocorrelation. In general, the closer is x_1 to x_3 (and x_2 to x_4) and the longer is a, the stronger should be the correlation. In fact when x_1 coincides with x_3 (and x_2 with x_4) we should have perfect correlation. The only question is what form the correlation takes as x_3 approaches x_2.

To answer this we consider the discrete analogue of equation (6.14):

$$Z(x + d) - \mu = \lambda_0 Y(x + d) + \lambda_1 Y(x + d + 1) + \lambda_2 Y(x + d + 2)$$
$$+ \cdots + \lambda_{a-1} Y(x + d + a - 1), \qquad (6.15)$$

where the $\lambda_0, \lambda_1, \ldots, \lambda_{a-1}$ are weights, here all equal to $1/a$, and $d = 1/2a$ is half the distance between two successive points in the sequence. All more distant members, say $Y(x + d + a - 1 + b)$, of the series carry zero weight. Suppose that $Y(x)$ is a white noise process; then $Z(x)$ is a moving average process of order $a - 1$. Further, if the variance of $Y(x)$ is σ_Y^2 then that of $Z(x)$ is

$$\sigma_Z^2 = \lambda_0^2 \sigma_Y^2 + \lambda_1^2 \sigma_Y^2 + \lambda_2^2 \sigma_Y^2 + \cdots + \lambda_{a-1}^2 \sigma_Y^2$$

$$= \sigma_Y^2 \sum_{i=0}^{a-1} \lambda_i \lambda_i$$

$$= \sigma_Y^2 / a, \tag{6.16}$$

which is familiar as the variance of a mean. It is also the covariance at lag 0, $C(0)$. We now want the covariances for the larger lags. These are obtained simply by extension from the above equation:

$$C(h) = \sigma_Y^2 \sum_{i=0}^{a-1-h} \lambda_i \lambda_{i+h} = \sigma_Y^2 \frac{a - h}{a^2}. \tag{6.17}$$

The covariances are in order, for $h = 0, 1, 2, \ldots, a - 1, a$,

$$\frac{a - 0}{a^2} \sigma_Y^2, \ \frac{a - 1}{a^2} \sigma_Y^2, \ \frac{a - 2}{a^2} \sigma_Y^2, \ldots, \ \frac{a - a + 1}{a^2} \sigma_Y^2, \ \frac{a - a}{a^2} \sigma_Y^2.$$

Dividing through by $C(0)$, we obtain the autocorrelations, $\rho(h)$, as

$$1, \ (a - 1)/a, \ (a - 2)/a, \ldots, \ (a - a + 1)/a, \ 0.$$

In words, the covariance and autocorrelation functions decay linearly with increasing h until $h = a$, at which point it is 0. Then the autocorrelation coefficient at any h is simply equal to the proportion of the filter that overlaps when the filter is translated by h. The variogram is obtained simply from relation (4.13) by

$$\gamma(h) = C(0) - C(h)$$

$$= \sigma_Y^2 \frac{a - h}{a^2} = \frac{\sigma^2}{a}\left(\frac{h}{a}\right) = c\left(\frac{h}{a}\right), \tag{6.18}$$

since $c = \sigma_Y^2 / a = C(0)$.

Circular model

The formula for the circular variogram is

$$\gamma(h) = \begin{cases} c\left\{1 - \dfrac{2}{\pi}\cos^{-1}\left(\dfrac{h}{a}\right) + \dfrac{2h}{\pi a}\sqrt{1 - \dfrac{h^2}{a^2}}\right\} & \text{for } h \leqslant a, \\ c & \text{for } h > a. \end{cases} \tag{6.19}$$

The parameters c and a are again the sill and range. The function curves tightly as it approaches the range (see Figure 6.3(b)), and its gradient at the origin is $4c/\pi a$. It is CNSD in $\mathbb{R}^1$ and $\mathbb{R}^2$, but not in $\mathbb{R}^3$.

This model can be derived in a way analogous to that of the bounded linear model from the area of intersection, A, of two discs of diameter a, the centres of which are separated by distance h. Matérn (1960) did this by considering the densities with which points are distributed at random by a Poisson process in two overlapping circles. This area is

$$A = \begin{cases} \dfrac{1}{2}a^2 \cos^{-1}\left(\dfrac{h}{a}\right) - \dfrac{h}{2\pi}\sqrt{a^2 - h^2} & \text{for } h \leqslant a, \\ 0 & \text{for } h > a. \end{cases} \qquad (6.20)$$

If we express this as a fraction of the area, $\pi a^2/4$, of one of the circles, in the same way as we expressed the fraction of the linear filter that overlapped along the line above, then we obtain the autocorrelation for the separation:

$$\rho(h) = \dfrac{2}{\pi}\left\{\cos^{-1}\left(\dfrac{h}{a}\right) - \dfrac{h}{a}\sqrt{1 - \dfrac{h^2}{a^2}}\right\} \qquad \text{for } h \leqslant a. \qquad (6.21)$$

Then from relation (4.13) the variogram, equation (6.19) above, follows.

Spherical model

By a similar line of reasoning we can derive the three-dimensional analogue of the circular model to obtain the spherical correlation function and variogram. The volume of intersection of two spheres of diameter a with their centres h apart is

$$V = \begin{cases} \dfrac{\pi}{4}c(\tfrac{2}{3}a^3 - a^2h + \tfrac{1}{3}h^3) & \text{for } h \leqslant a, \\ 0 & \text{otherwise.} \end{cases} \qquad (6.22)$$

The volume of a sphere is $\tfrac{1}{6}\pi a^3$, and so dividing by it gives the autocorrelation

$$\rho(h) = \begin{cases} 1 - \dfrac{3h}{2a} + \dfrac{1}{2}\left(\dfrac{h}{a}\right)^3 & \text{for } h \leqslant a, \\ 0 & \text{for } h > a, \end{cases} \qquad (6.23)$$

and the variogram is

$$y(h) = \begin{cases} c\left\{\dfrac{3h}{2a} - \dfrac{1}{2}\left(\dfrac{h}{a}\right)^3\right\} & \text{for } h \leqslant a, \\ c & \text{for } h > a. \end{cases} \qquad (6.24)$$

The spherical model seems the obvious one to describe variation in three-dimensional bodies of rock, and it has proved well suited to them. It would seem less obviously suited for describing the variation in one and two dimensions, which is usually what is needed in soil and land resource survey. Yet it nearly always fits experimental results from soil sampling better than the one- and two-dimensional analogues. The function curves more gradually than they do (Figure 6.3(c)), and the reason is probably that there are additional sources of variation at other scales that it can represent. Its gradient at the origin is $3c/2a$. It is CNSD in $\mathbb{R}^2$ and $\mathbb{R}^1$ as well as in $\mathbb{R}^3$.

The spherical function is one of the most frequently used models in geostatistics, in one, two and three dimensions. It represents transition features that have a common extent and which appear as patches, some with large values and others with small ones. The average diameter of the patches is represented by the range of the model. This interpretation can be seen by simulating a large field of values using the function as the generator. Figures 6.4 and 6.5(a) are examples in which values have been simulated on a 256×256 square grid with unit interval. The model had a sill variance, $c = 1.0$, and ranges of $a = 15$, 25 and 50 units in Figures 6.4(a), 6.4(b) and 6.5(a), respectively.

Pentaspherical model

Following Matérn (1960), McBratney and Webster (1986) extended the line of reasoning to obtain the five-dimensional analogue of the above, the pentaspherical function:

$$y(h) = \begin{cases} c\left\{\dfrac{15}{8}\dfrac{h}{a} - \dfrac{5}{4}\left(\dfrac{h}{a}\right)^3 + \dfrac{3}{8}\left(\dfrac{h}{a}\right)^5\right\} & \text{for } h \leqslant a, \\ c & \text{for } h > a. \end{cases} \qquad (6.25)$$

It is useful in that its curve is somewhat more gradual than that of the spherical model (Figure 6.3(d)). Its gradient at the origin is $15c/8a$. Again it is CNSD in $\mathbb{R}^1$, $\mathbb{R}^2$ and $\mathbb{R}^3$.

Exponential model

A function that is also much used in geostatistics is the negative exponential:

$$y(h) = c\left\{1 - \exp\left(-\dfrac{h}{r}\right)\right\}, \qquad (6.26)$$

with sill c, and a distance parameter, r, that defines the spatial extent of the model. The function approaches its sill asymptotically, and so it does not have a finite range. Nevertheless, for practical purposes it is convenient to assign it an effective range, and this usually taken as the

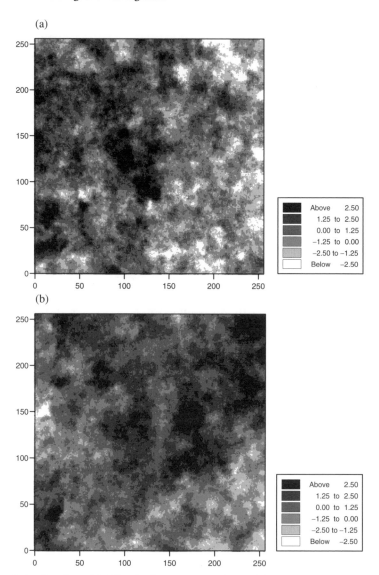

Figure 6.4 Simulated fields of values using spherical functions (equation (6.24)), with distance parameters (a) $a = 15$, (b) $a = 25$.

distance at which γ equals 95% of the sill variance, approximately $3r$. Its slope at the origin is c/r. Figure 6.6(a) shows it.

The function has an important place in statistical theory. It represents the essence of randomness in space. It is the variogram of first-order autoregressive and Markov processes. Its equivalent autocorrelation function has been the basis of several theoretical studies of the efficiency

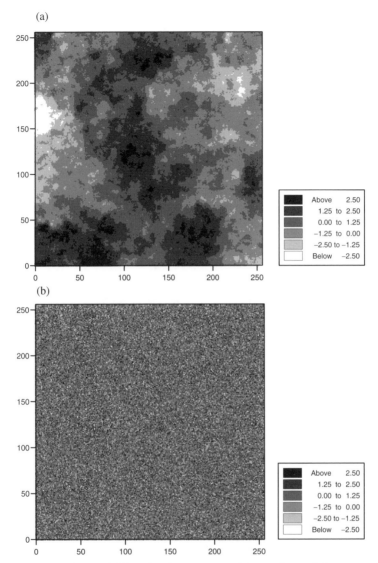

Figure 6.5 Simulated fields of values using: (a) a spherical function (equation (6.24)), with distance parameter $a = 50$; and (b) a pure nugget variogram (equation (6.31)).

sampling designs by, for example, Cochran (1946), Yates (1948), Quenouille (1949) and Matérn (1960). We should expect variograms of this form where differences in soil type are the main contributors to soil variation and where the boundaries between types occur at random as a Poisson process. Burgess and Webster (1984) found this to be the situation in

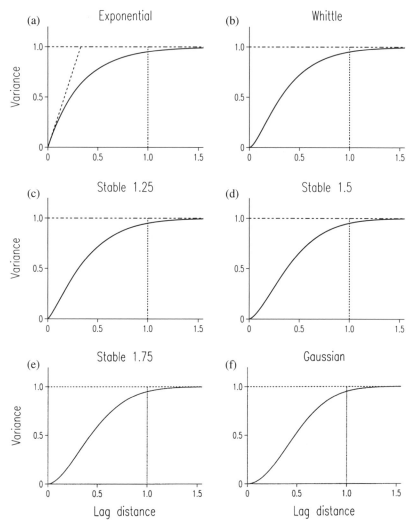

Figure 6.6 Models with asymptotic bounds. All are scaled so that the effective range where the function reaches 0.95 of its sill is approximately 1, marked by the vertical lines on the graphs. (a) $\alpha = 1$ (exponential), $r = 0.333$; (b) Whittle, $r = 0.25$; (c) $\alpha = 1.25$ (stable), $r = 0.416$; (d) $\alpha = 1.5$ (stable), $r = 0.478$; (e) $\alpha = 1.75$ (stable), $r = 0.533$; (f) $\alpha = 2$ (Gaussian), $r = 1/\sqrt{3}$.

many instances. If the intensity of the process is η then the mean distance between boundaries is $\bar{d} = 1/\eta$, and the variogram is

$$y(h) = c\{1 - \exp(-h/\bar{d})\}$$
$$= c\{1 - \exp(-\eta h)\}. \tag{6.27}$$

Put another way, this is the variogram of a transition process in which the structures have random extents.

The simulated fields obtained using an exponential function with an asymptote approaching 1.0 and distance parameters, r, of 5 and 16 are shown in Figure 6.7(a) and 6.7(b), respectively. The patches of large and small values in the two fields are similarly irregular, but the average sizes of the patches show the different spatial scales of the generator.

Whittle's elementary correlation

Whittle (1954) showed that a simple stochastic diffusion process also has an exponential variogram in one and three dimensions. In $\mathbb{R}^2$, however, the process leads to Whittle's *elementary correlation*, given by

$$\gamma(h) = c\left\{1 - \frac{h}{r}K_1\left(\frac{h}{r}\right)\right\}. \tag{6.28}$$

The parameter c is the sill, as before, the *a priori* variance of the process, r is a distance parameter, and K_1 is the modified Bessel function of the second kind. Like the exponential function, Whittle's function approaches its sill asymptotically and so has no definite range. Its effective range may be chosen as for the exponential function where the semivariance reaches 95% of the sill, and this is at approximately $4r$. The function approaches the origin with a decreasing gradient, however, and appears slightly sigmoid when plotted (Figure 6.6(b)).

Gaussian model

Another function with reverse curvature near the origin recurs again and again in geostatistical texts and software packages. Its graph is shown in Figure 6.6(f). It is the so-called Gaussian model with equation

$$\gamma(h) = c\left\{1 - \exp\left(-\frac{h^2}{r^2}\right)\right\}. \tag{6.29}$$

Once more, c is the sill and r is a distance parameter. The function approaches its sill asymptotically, and it can be regarded as having an effective range of approximately $\sqrt{3}r$, where it reaches 95% of its sill variance.

A serious disadvantage of the model is that it approaches the origin with zero gradient, which we saw above as the limit for random variation and at which the underlying variation becomes continuous and twice differentiable. This can lead to unstable kriging equations (see below) and bizarre effects when used for estimation—see Wackernagel (1995) for examples.

In general we deprecate this model. If a variogram appears somewhat sigmoid then we recommend the theoretically attractive Whittle function. Alternatively, if the reverse curvature is stronger you may replace the

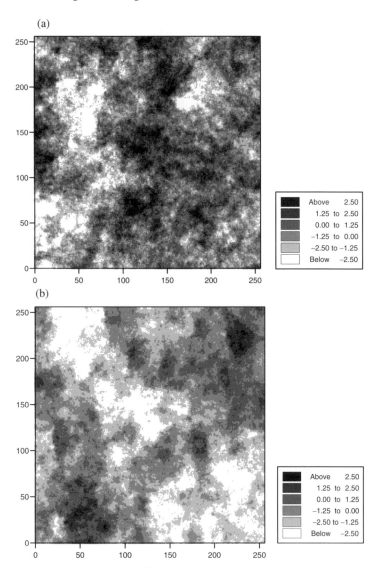

Figure 6.7 Simulated fields of values using exponential functions (equation (6.26)), with distance parameters (a) $r = 5$, (b) $r = 16$.

exponent 2 in equation (6.29) by an additional parameter, say α, with a value less than 2:

$$\gamma(h) = c\left\{1 - \exp\left(-\frac{h^\alpha}{r^\alpha}\right)\right\}. \tag{6.30}$$

Wackernagel (1995) calls these 'stable models'. Some examples of them are shown in Figure 6.6(c)-(e) with various values of α.

We do not know of any application of this model in an environmental survey.

There are other simple models, some of which are favoured in particular disciplines such as geophysics because they have theoretical bases in those fields. If you work in such a special field then you should ask whether there are preferred functions for the particular applications.

Pure nugget

Although the limiting value 0 of the exponent of equation (6.10) for the power function was excluded because it would give a constant variance, we do need some way of expressing such a constant because that is what appears in practice. We do so by defining a 'pure nugget' variogram as

$$\gamma(h) = c_0\{1 - \delta(h)\}, \tag{6.31}$$

where c_0 is the variance of the process, and $\delta(h)$ is the Kronecker δ which takes the value 1 when $h = 0$ and is zero otherwise. If the variable is continuous, as almost all properties of the soil and natural environment are, then a variogram that appears as pure nugget has almost certainly failed to detect the spatially correlated variation because the sampling interval was greater than the scale of spatial variation.

Since the nugget variance is constant for all $\mathbf{h}$, $|\mathbf{h}| > 0$, it is usually denoted simply by the variance c_0. Figure 6.5(b) shows the simulated field from a pure nugget variogram. There is no detectable pattern in the variation as there is in Figures 6.4, 6.5(a) and 6.7.

6.3 COMBINING MODELS

As is apparent in Figures 6.3 and 6.6, all the above functions have simple shapes. In many instances, however, especially where we have many data, variograms appear more complex, and we may therefore seek more complex functions to describe them. The best way to do this is to combine two or more simple models. Any combination of CNSD functions is itself CNSD. Do not look for complex mathematical solutions the properties of which are unknown.

The most common requirement is for a model that has a nugget component in addition to an increasing, or structured, portion. So, for example, the equation for an exponential variogram with a nugget may be written as:

$$\gamma(h) = c_0 + c\left\{1 - \exp\left(-\frac{h}{r}\right)\right\}, \tag{6.32}$$

and an example is shown Figure 6.8(a). Figure 6.9 shows the simulated fields for an exponential variogram with parameters $c_0 = 0.333, c = 0.667$

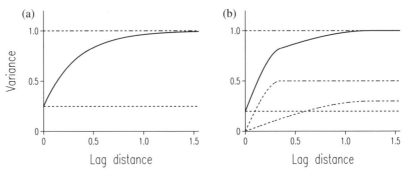

Figure 6.8 Combined (nested) models: (a) single exponential with sill 0.75 plus a nugget variance of 0.25; (b) double spherical with ranges 0.35 and 1.25 and corresponding sills 0.3 and 0.5 plus a nugget variance of 0.2 with the components shown separately.

and distance parameters, r, of 5 and 16 as before. The speckled appearance within the patches is the result of the nugget variance.

Spatial dependence may occur at two distinct scales, and these may be represented in the variogram as two spatial components. The nested spherical, or double spherical, function is the one that has been used most often in these circumstances. Its equation is

$$
\gamma(h) = \begin{cases}
c_1 \left\{ \dfrac{3h}{2a_1} - \dfrac{1}{2}\left(\dfrac{h}{a_1}\right)^3 \right\} + c_2 \left\{ \dfrac{3h}{2a_2} - \dfrac{1}{2}\left(\dfrac{h}{a_2}\right)^3 \right\} & \text{for } 0 < h \leqslant a_1, \\[3mm]
c_1 + c_2 \left\{ \dfrac{3h}{2a_2} - \dfrac{1}{2}\left(\dfrac{h}{a_2}\right)^3 \right\} & \text{for } a_1 < h \leqslant a_2, \\[3mm]
c_1 + c_2 & \text{for } h > a_2,
\end{cases}
$$
(6.33)

where c_1 and a_1 are the sill and range of the short-range component of the variation, and c_2 and a_2 are the sill and range of the long-range component. If it appears to need a nugget then that can be added as a third component, and Figure 6.8(b) shows this combination.

6.4 PERIODICITY

A variogram may seem to fluctuate more or less periodically, rather than increase monotonically, and we might try to describe it with a periodic function. The simplest such function is a sine wave, see Figure 6.10(a), with equation

$$
\gamma(h) = W \left\{ 1 - \cos\left(\frac{2\pi h}{\omega}\right) \right\},
$$
(6.34)

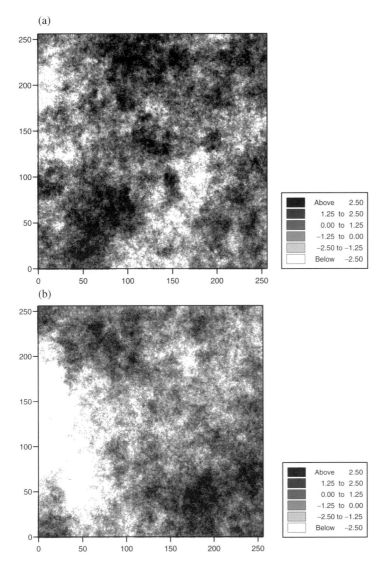

Figure 6.9 Simulated fields of values using exponential functions with nugget variance one-third of the total (equation (6.32)): (a) with distance parameter, $r = 5$; (b) with $r = 16$.

where W and ω are the amplitude and length of the wave, respectively.

The gradient at the origin is 0, which, as mentioned above, is undesirable. Usually, however, we find that the periodicity is superimposed on some other source of variation and that the combined model increases from the origin more steeply. Figure 6.10(b) shows an example of it super-

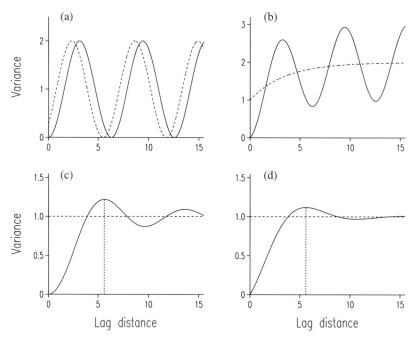

Figure 6.10 Periodic and hole effect models: (a) simple sine wave of length 2π (solid) and with phase shift (dashed line); (b) sine wave superimposed on an exponential function; (c) damped sine wave of period 2.5π (with maximum marked at $h \approx 5.6$); and (d) model with Bessel function J_0 (equation (6.42)), with distance parameter set to 10 (giving maximum at $h \approx 5.6$).

imposed on an exponential function. An example from actual soil survey is illustrated in the next chapter.

We may also find that we need to move the curve along the abscissa so that it increases more nearly linearly from lag 0. Such a function appears as the dashed line in Figure 6.10(a). In other words, we may introduce a phase shift, ϕ. If we designate the angle $2\pi h / \omega$ as θ for simplicity, then the equation becomes

$$\gamma(h) = W\{1 - \cos(\theta - \phi)\}. \tag{6.35}$$

These functions are valid for one dimension only; they are not CNSD in $\mathbb{R}^2$ and $\mathbb{R}^3$. In two and three dimensions the fluctuation must damp, i.e. become less pronounced with increasing lag distance. Damping can be achieved by dividing the sine or cosine function by a function of the lag distance. Choosing again the simplest function, we can write

$$\gamma(h) = W\left(1 - \frac{1}{\theta}\sin\theta\right), \tag{6.36}$$

which increases from zero at $h = 0$, and this appears in Figure 6.10(c). Again, one may add a phase shift, ϕ.

This model is valid in one, two and three dimensions. Journel and Huijbregts (1978) show that it has a relative amplitude, which they define as

$$\beta = \frac{1}{c}\{\max[y(h)] - c\},\qquad(6.37)$$

where $\max[y(h)]$ is the maximum semivariance of the function, and c is the *a priori* variance, the horizontal line in Figure 6.10(c). In equation (6.36) β is approximately 0.217, and it occurs where $2.5\pi h/\omega \approx 5.6$. It is the maximum for a periodic model in $\mathbb{R}^3$.

The general equation (6.35) with two non-linear parameters, ω and ϕ, is somewhat difficult to handle computationally, and it is often converted to one that is easier. Again replacing $2\pi h/\omega$ by θ, then from trigonometry we can write the equation as

$$y(h) = W\cos\theta\cos\phi + W\sin\theta\sin\phi.\qquad(6.38)$$

The quantities $W\cos\phi$ and $W\sin\phi$ are designated c_1 and c_2, and the whole equation becomes

$$y(h) = c_1\cos\theta + c_2\sin\theta.\qquad(6.39)$$

Thus c_1 and c_2 are now linear parameters that determine the amplitude and phase of the periodicity. If the equation is given in this form then the phase and amplitude, respectively, are readily obtained from the relations

$$\begin{aligned}\phi &= \arctan(c_2/c_1),\\ W &= c_1/\cos\phi.\end{aligned}\qquad(6.40)$$

The damped version of equation (6.39) is

$$y(h) = (c_1\cos\theta + c_2\sin\theta)/\theta.\qquad(6.41)$$

As above, it is not necessarily CNSD: it depends on the values of c_1, c_2 and ω, and it must be tested for validity with the particular coefficients.

Usually in such cases the damping is such that only the first undulation is substantial. The corresponding covariance function, see Figure 4.3(f), appears to have a single depression in it: it is said to exhibit a *hole effect*.

Another model that might describe a less pronounced hole effect satisfactorily embodies the Bessel function J_0:

$$y(h) = c\left\{1 - \exp(-h/r)J_0\left(\frac{2\pi h}{\omega_J}\right)\right\},\qquad(6.42)$$

where J_0 is the Bessel function of the first kind, and ω_J is a distance parameter corresponding roughly to wavelength. The maximum is only 0.118 times the variance of the process. Figure 6.10(d) shows an example in which ω_J has been set to 10 to give a maximum at the same lag distance, 5.6, as in the truly periodic models.

Practitioners should treat wavy experimental variograms with caution. The experimental values are themselves correlated, the more so as the correlation in the original data strengthens. One consequence of this is that any underlying wave-like fluctuation tends to be exaggerated in the estimates: the wave does not damp as much as you might expect, even with moderately long runs (200–300) of data. Before trying to fit a periodic function to such a set of points, the user should ask what evidence there is of periodicity in the phenomenon being investigated. If there is none and the apparent periodicity or hole is weak then do not try to force a periodic model on the variogram. This is a specific case of the more general advice that any variogram model should accord with what you know of the underlying variable, such as the soil, geology, landscape, or sources of pollution that you are studying.

6.5 ANISOTROPY

Variation can itself vary with direction. If it can be made to seem isotropic by transforming the horizontal scales, then it is called *geometric* or *affine* anisotropy. Such anisotropy can be taken into account by a simple linear transformation of the rectangular coordinates. It is perhaps best envisaged for a process with a spherical variogram in which the range, instead of being a constant, describes an ellipse in the plane of the lag. This is shown in Figure 6.11, where A is the maximum diameter of the ellipse, i.e. the range in the direction of greatest continuity (least change with separating distance), and B is the minimum diameter, perpendicular to the first,

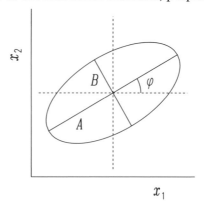

Figure 6.11 A representation of geometric anisotropy in which the ellipse describes the range of a spherical variogram in two dimensions. The diameter A is the maximum range of the model, B is the minimum range, and ϕ is the direction of the maximum range.

and is the range in the direction of least continuity (greatest change with separating distance). The angle ϕ is the direction in which the continuity is greatest. The equation for transformation is then

$$\Omega(\theta) = \{A^2 \cos^2(\theta - \phi) + B^2 \sin^2(\theta - \phi)\}^{1/2}, \qquad (6.43)$$

where Ω defines the anisotropy, and θ is the direction of the lag.

If we insert $\Omega(\theta)$ into the spherical function then we have

$$\gamma(h, \theta) = \begin{cases} c\left\{\dfrac{3|\mathbf{h}|}{2\Omega(\theta)} - \dfrac{1}{2}\left(\dfrac{|\mathbf{h}|}{\Omega(\theta)}\right)^3\right\} & \text{for } 0 < |\mathbf{h}| \leqslant \Omega(\theta), \\ c & \text{for } |\mathbf{h}| > \Omega(\theta). \end{cases} \qquad (6.44)$$

We can derive from equation (6.43) an anisotropy ratio, $R = A/B$, and it may be convenient to rewrite equation (6.43) with this replacing B:

$$\Omega(\theta) = \left\{A^2 \cos^2(\theta - \phi) + \left(\frac{A}{R}\right)^2 \sin^2(\theta - \phi)\right\}^{1/2}. \qquad (6.45)$$

This transformation in effect allows the variation to be represented in an isotropic form. It is as if the soil were on a rubber sheet and stretched in the direction parallel to B until $B = A$ and the ellipse becomes a circle.

The function $\Omega(\theta)$ can be applied to the power function for unbounded variation:

$$\begin{aligned} \gamma(h, \theta) &= \{A^2 \cos^2(\theta - \phi) + B^2 \sin^2(\theta - \phi)\}^{1/2}|\mathbf{h}|^\alpha \\ &= \Omega(\theta)|\mathbf{h}|^\alpha. \end{aligned} \qquad (6.46)$$

Here the roles of A and B are inverted; they are now gradients, with A, the larger, being the gradient in the direction of greatest rate of change and B, the smaller, being the gradient in the direction of the smallest rate.

6.6 FITTING MODELS

The models described above are those that are commonly used for variograms in resource survey. All are theoretically based. Our task now is to fit them to the experimental or sample values. One might have thought after so many years of geostatistical development and practice—more than 30 since Matheron (1965) published his seminal thesis and more than 20 since the first textbooks appeared—that the task would be straightforward, with standard algorithms and well-tried software. If so, one would be wrong. Choosing models and fitting them to data remain among the most controversial topics in geostatistics.

There are still practitioners who fit models by eye and who defend their practice with vigour. They may justify their attitude on the grounds that when kriging the resulting estimates are much the same for all reasonable models of the variogram—so why worry about refinement? There are others who fit models numerically and automatically using 'black-box' software, often without any choice, judgement or control. This too can have unfortunate consequences. However, there is controversy among those who fit models mathematically about which methods to use and by what criteria they should judge success.

Fitting models is difficult for several reasons, including the following:

(i) The accuracy of the observed semivariances is not constant.

(ii) The variation may be anisotropic.

(iii) The experimental variogram may contain much point-to-point fluctuation.

(iv) Most models are non-linear in one or more parameters.

Items (i)–(iii) make fitting by eye unreliable. The first two impair one's intuition, (i) because the brain cannot judge the weights to attribute to the semivariances, and (ii) because one cannot see the variogram in three dimensions without constructing a stereogram or physical model, and for three-dimensional variation one needs a fourth dimension. Scatter, item (iii), usually means that any one of several models might be drawn through the values. It can also lead to unstable mathematical solutions, and it exacerbates the consequences of item (iv) because the non-linear parameters must be found by iteration. Further, at the end one should be able to put standard errors on the estimates of the parameters.

We also warn against the practice, still common, of choosing the dispersion variance in a finite region to estimate the sill of a bounded model for the regional variogram. For such a region the sill is always greater than the dispersion variance. Their relation is shown in Figure 4.4. The curve is the variogram of a second-order stationary process in one dimension of finite length, as on a transect. The variogram is extended to the limit of the transect, and in these circumstances the two shaded portions of the graph should be equal. Clearly the sill, the *a priori* variance of the process, must exceed the dispersion variance, which is estimated by the variance of the data.

We recommend a procedure that embodies both visual inspection and statistical fitting, as follows. First plot the experimental variogram. Then choose, from the models listed above, one or more with approximately the right shape and with sufficient detail to honour the principal trends in the experimental values that you wish to represent. Then fit each model in turn by weighted least squares, i.e. by minimizing the sums of squares, suitably weighted (see below), between the experimental and fitted values. Lastly, inspect the result graphically by plotting the fitted model on the same pair of axes as the experimental variogram. Does the fitted function look reasonable? If all the plausible models seem to fit well you might

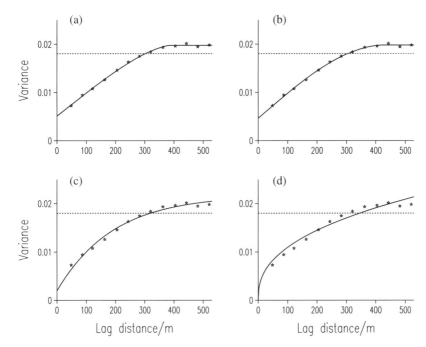

Figure 6.12 Experimental variogram of $\log_{10}$ K at Broom's Barn Farm with plausible models fitted to it: (a) circular; (b) spherical; (c) exponential; (d) power function.

choose, finally, from among them the one with smallest residual sum of squares or smallest mean square.

The data on exchangeable potassium at Broom's Barn Farm illustrate the procedure. The experimental variogram computed at intervals of 40 m appears in Figure 6.12 in its isotropic form. Its form is simple; it increases smoothly in a gentle curve from near the origin and seems to flatten near the maximum lag to which it has been computed. Plausible models to describe it are the power function, the circular function, the exponential, and the spherical function. These have been fitted, and they too appear in Figure 6.12. The power function evidently does not fit as well as the others. But which of the last three is the best? Table 6.1 shows that it is the spherical function that has the smallest mean squared residual (MSR).

Never accept a fit without inspecting it afterwards; it might be poor because

(i) you chose an unsuitable model in the first place;

(ii) you gave poor estimates of the parameters at the start of the iteration;

(iii) there was lot of scatter in the experimental variogram; or

Table 6.1 Models fitted to the variogram of $\log_{10} K$ at Broom's Barn Farm, their parameter values, and the mean squared residual (MSR). The symbols are as defined in the text.

Model	c_0	c	a (m)	r (m)	w	α	MSR
Circular	0.005 12	0.014 62	386.6				0.000 172
Spherical	0.004 66	0.015 15	432.0				0.000 155
Pentaspherical	0.004 21	0.015 70	514.1				0.000 248
Exponential	0.001 96	0.019 73		190.4			0.001 054
Power function	0				0.001 73	0.400	0.003 295

(iv) the computer program was faulty.

Further, bear in mind the advice above, namely, that the model should accord with what you know of the region.

Fitting models in this way is a form of non-linear regression, and you might think of writing your own program to do it. We recommend that unless you are proficient in numerical analysis you do not. There are now several well-tried programs written by professionals that fit models by weighted least squares. These include Genstat (Genstat 5 Committee, 1993), in which the standard models listed above are already programmed (Genstat 5 Committee, 1995), and MLP (Ross, 1987), both of which we use, and SAS (SAS Institute, 1985). The last uses the Levenberg–Marquardt method, which has almost become a standard for non-linear model fitting (Marquardt, 1963). We give an example of a Genstat program for fitting non-linear models in Appendix B. If you do not have access to any of these programs then you might take the code for the Marquardt algorithm in Fortran from Press *et al.* (1992). Ratkowsky (1983) also tackles the subject in a clear and practical way, and the book includes a suite of subroutines for modelling.

We can call the above approach 'fit statistically, view afterwards'. Another approach is the reverse: 'fit visually, statistics afterwards'. Pannatier (1995) takes this route with his program Variowin, which is interactive in a Windows environment. In Variowin you form the experimental variogram from sample data and you display it on the computer's screen. You select a plausible model from those embodied in the program—there are rather few—and give starting values for its parameters from which the machine draws a graph. The program simultaneously computes a goodness-of-fit criterion, which is a standardized residual sum of squares. You then adjust the values of the parameters to try to improve the fit visually, and as you do so the program redraws the model in real time and recomputes the goodness-of-fit criterion. It also compares the criterion with the best it has found to date and stores the criterion's value and the associated values of the parameters if the new fit is better. You terminate the fitting when you are satisfied with the approximation or no further improvement seems possible. In our experience it works well, though never better than Genstat or MLP (Webster and Oliver, 1997).

6.6.1   What weights?

We mentioned above that the experimental semivariances, $\hat{y}(\mathbf{h})$, vary in their reliability, partly because they are based on varying numbers of paired comparisons, $m(\mathbf{h})$ in equation (5.4), and partly because the confidence in an estimate of variance decreases as the variance increases. In general, therefore, assigning equal weight to all $\hat{y}(\mathbf{h})$ is unsatisfactory, especially if the $m(\mathbf{h})$ vary widely with changing $\mathbf{h}$. We can take the latter into account simply by weighting in proportion m. The inverse relation between the reliability of an estimate of variance and the variance itself led Cressie (1985) to propose a more elaborate weight at a lag $\mathbf{h}_j$ in the form

$$m(\mathbf{h}_j)/y^{*2}(\mathbf{h}_j),$$

where $y^{*2}(\mathbf{h}_j)$ is the value of semivariance predicted by the model. Mc-Bratney and Webster (1986) refined this further as

$$m(\mathbf{h}_j)\hat{y}(\mathbf{h}_j)/y^{*3}(\mathbf{h}_j),$$

where $\hat{y}(\mathbf{h}_j)$ is the observed value of the semivariance at $\mathbf{h}_j$. Both of the last two schemes tend to give more weight at the shorter lags than does weighting on the numbers of pairs alone, and so the fitting is closer there. This is usually desirable for kriging (see Chapter 8), though it might be less desirable if the aim is to estimate the spatial scale of variation.

The process of fitting must iterate even where all the parameters are linear because the weights in the two schemes depend on the values expected from the model. Our experience is that in most instances there is little change after the first iteration, which is therefore enough.

6.6.2   How complex?

Let us return to the question we posed at the beginning of the chapter: how closely should the model follow the fluctuation in the experimental variogram? The best simple model, with few parameters, might fit the experimental variogram poorly, especially if there is much point-to-point scatter. We might seek a more complex model, therefore, bearing in mind that it is almost always possible to improve the fit in the least-squares sense by increasing the numbers of parameters, say p. We could continue to increase p until the model fitted perfectly, but clearly that is not a sensible answer. We must compromise between parsimony (few parameters) and close fit (more parameters), and one way of achieving that is to use Akaike's (1973) information criterion (AIC):

$$AIC = -2 \ln(\text{maximized likelihood}) + 2(\text{number of parameters}).$$

The AIC is estimated by

$$\widehat{AIC} = \left\{ n \ln\left(\frac{2\pi}{n}\right) + n + 2 \right\} + n \ln R + 2p, \qquad (6.47)$$

where n is the number of points on the variogram, p is the number of parameters in the model, and R is the mean of the squared residuals between the experimental values and the fitted model. We may then choose the model for which $\widehat{AIC}$ is least. The quantity in braces is constant for any one experimental variogram, so we need compute only

$$\hat{A} = n \ln R + 2p. \qquad (6.48)$$

Least-squares fitting minimizes R, but if R is further diminished only by increasing p (n is constant) then there is a penalty, and it might be too big.

We illustrate the application of the AIC to modelling the variogram of available copper in the topsoil of the Borders Region of Scotland. The data are from an original study by McBratney *et al.* (1982). There were some 2000 values from the eastern portion of the Region. They were transformed to their common logarithms to stabilize their variances, and an isotropic experimental variogram was computed. It appears as the plotted points in Figure 6.13. Any smooth curve through the points will have an intercept, so we include a nugget variance in the model. By fitting single spherical and exponential functions, with weights proportional to the numbers of pairs, we obtain the curves of best fit shown in Figures 6.13(a) and 6.13(b), respectively. Clearly both fit poorly near the ordinate. The values of the parameters, the residual sum of squares and $\hat{A}$ are listed in Table 6.2, from which it is evident that the exponential function is the better. If we add another spherical or exponential component we obtain the more detailed curves in Figures 6.13(c) and 6.13(d), respectively. Now the double spherical is evidently the best, with the smallest mean squared residual. It also has the smallest $\hat{A}$, and so in this instance we choose this more elaborate model.

This solution is valid for weighted least-squares fitting provided the weights remain constant, as when they are simply set in proportion to the numbers of paired comparisons.

Webster and McBratney (1989) discuss the AIC in some detail, show its equivalence to an F test for nested models, and suggest other possible criteria.

Another method for judging the goodness of a model is *cross-validation*. This involves comparing kriged estimates and their variances, and we defer it until we have described kriging.

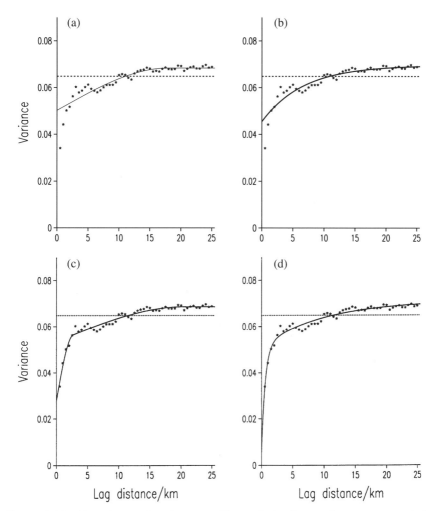

Figure 6.13   Experimental variogram of $\log_{10}$ Cu in the Borders Region with fitted models: (a) single spherical; (b) exponential; (c) double (nested) spherical; (d) double exponential.

Table 6.2 Models fitted to the variogram of $\log_{10}$ Cu in the Borders Region, estimates of their parameters, the mean squared residual (MSR), and the variable part of the Akaike information criterion ($\hat{A}$). The symbols are as defined in the text.

Model	Sills			Distance parameters (km)				MSR	$\hat{A}$
	c_0	c_1	c_2	a_1	a_2	r_1	r_2		
Spherical	0.050 27	0.018 05		18.0				0.068 22	−128.3
Exponential	0.045 49	0.024 03				6.65		0.060 46	−134.3
Double spherical	0.027 67	0.025 85	0.015 05	2.7	20.5			0.029 94	−165.4
Double exponential	0.005 67	0.045 66	0.019 75			0.59	9.99	0.036 16	−155.7

7

Spectral analysis

In some places the land varies laterally in a regular fashion. The most obvious regular patterns are man-made. They include the characteristic ridge-and-furrow of the English clay lands, and orchards and plantations in which fruit trees and other crops are arranged in lines with constant intervals between them. The dynamic properties of the soil are likely to vary in tune with them and so also have a regularity. Forest is established on peaty soil by planting young trees on the upturned sod after deep ploughing in lines. Less obvious are the long-lasting patterns of former ploughing on crop yield, revealed by McBratney and Webster (1981), and the effects of earlier drainage schemes on the present-day soil described by Burrough *et al.* (1985). In all these the regularity is deliberate.

Natural features may also seem regular. Examples are the frost polygons of the Arctic region and their fossil relics in the Northern temperate zone (e.g. Hodge and Seale, 1966), the patterns of termite mounds in Africa, especially evident on some of the East African plains (e.g. Scott *et al.*, 1971) and in the Miombo woodland of Zambia and Congo (Zaïre) and the gilgai of Australia (e.g. Hallsworth *et al.*, 1955; Webster, 1977).

The experimental variograms of such patterns fluctuate with evident periodicity, as Webster (1977) discovered. Other periodic patterns can arise from cultivation and land management (McBratney and Webster, 1981; Burrough *et al.*, 1985). The previous chapter mentioned basic periodic functions that might be used to describe the fluctuation, but deferred illustration until now so that we can deal with it and spectral analysis together.

7.1 LINEAR SEQUENCES

More often than not we encounter periodicity in linear, i.e. one-dimensional, sequences of data comprising records made at regular intervals in either time or space (see, for example, Oliver *et al.*, 1997). Spatial examples include:

- photographic and radiometric survey by aircraft;

- bathymetric and sonar survey from ships;
- electric logs of boreholes for oil exploration;
- pollen counts through peat;
- isotope measurements through polar ice;
- transect surveys of soil.

In some instances each line is one of several or many in $\mathbb{R}^2$ or $\mathbb{R}^3$. In others the lines are isolated representatives of two-dimensional scenes. Variables, such as temperature, may also be recorded at regular intervals in time, and in that instance there is only one dimension. We can analyse the data by all of the standard geostatistical methods described above. However, if there is periodicity then it is often profitable to express the variation in relation to frequency rather than space or time, and this takes us into the realm of spectral analysis.

7.2 GILGAI TRANSECT

To illustrate an analysis of periodic variation we use the data from a survey by Webster (1977) of salinity on the Bland Plain of eastern Australia. This virtually flat plain is part of the Murray–Darling Basin. Its soil is dominantly clay, but with a more sandy surface horizon of variable thickness, alkaline and locally saline. One of its most remarkable features is its patterns of gilgai. The gilgais are small almost circular depressions from a few centimetres to as much as 50 cm deep in the plain and several metres across. The soil in their bottoms is usually clay and wetter than that elsewhere.

A paddock, at Caragabal, was surveyed by sampling at regular 4 m intervals along a transect almost 1.5 km long. At each of 365 sampling points a core of soil, 75 mm in diameter, was taken to 1 m, and segments of it were analysed in the laboratory. For present purposes we shall concern ourselves with just one variable, the electrical conductivity at 30–40 cm. Table 7.1 summarizes the data, which were strongly skewed and therefore transformed to logarithms for further analysis. Figure 7.1 shows the logarithm of conductivity plotted against position as the fine line. The bold line is a smoothing spline fitted to the data to filter out the short-range variation and reveal a fluctuation of longer range that appears regular.

The experimental variogram of the data is shown in Figure 7.2 as the plotted points, to which we have fitted a model with a periodic component. The full model is given by

$$y(h) = c_0 + wh + c\{\text{sph}(a)\} + c_1 \cos\left(\frac{2\pi h}{\omega}\right) + c_2 \sin\left(\frac{2\pi h}{\omega}\right). \quad (7.1)$$

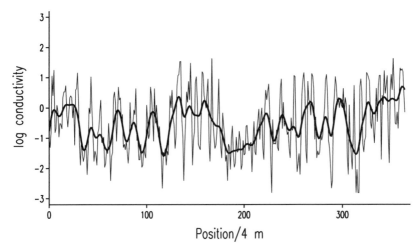

Figure 7.1 Trace of common logarithm of electrical conductivity at Caragabal (fine line) with a smoothing spline (bold) added as an aid to see the suspected periodicity.

Table 7.1 Summary statistics of electrical conductivity in the soil at 30-40 cm at Caragabal.

	Electrical conductivity	
	mS cm^{-1}	log$_{10}$(mS cm^{-1})
Minimum	0.06	−1.214
Maximum	5.10	0.707
Mean	0.958	−0.229 8
Median	0.54	−0.266 8
Variance	0.959 48	0.192 05
Standard deviation	0.975	0.438
Skewness	1.642	0.101

It comprises a small nugget, c_0, a linear component, wh, a spherical function with a short range, $c\{\text{sph}(a)\}$, and a sine wave, $c_1 \cos(2\pi h/\omega) + c_2 \sin(2\pi h/\omega)$. The values of these parameters are given in Table 7.2. The spherical component contributes most to the variance, with a sill of approximately 0.15 log(mS cm^{-1})^2. It represents repetitive variation of a kind that is not periodic. For present purposes the periodic component, though representing less of the variance with an amplitude of only 0.012, is of most interest. Its wavelength is 8.67 sampling units or 35 m. This is approximately equal to the average distance between the centres of the gilgai on the transect. It has a phase shift of -0.43 rad (about −25°). The linear component has only a very gentle gradient; it is of little practical

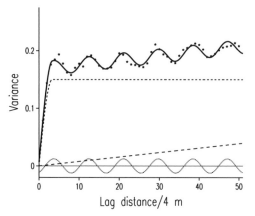

Figure 7.2   Variogram of log electrical conductivity. The points are the sample values, the heavy line is the fitted model comprising the four components shown with the lighter lines. The parameter values are listed in Table 7.2.

Table 7.2   Parameter values of model fitted to variogram of log electrical conductivity in the soil at 30–40 cm at Caragabal. Distances are in sampling intervals of 4 m, and angles are in radians.

Component	Parameter	Value
Nugget	constant, c_0	0.017 60
Linear	gradient, w	0.000 772
Spherical	sill, c	0.149 8
	range, a	3.323
Periodic	amplitude, W,	0.012 30
	wavelength, ω	8.667
	phase, ϕ	−0.435
	c_1	−0.011 16
	c_2	0.005 181

consequence, and we may regard the underlying variation as second-order stationary. The nugget variance is also very small. In passing, we note that the periodic component does not damp, and so the model is valid in one dimension only.

7.3   POWER SPECTRA

Let us now consider how to examine this variation in the frequency domain. We start by assuming that the underlying variable, $Z(\mathbf{x})$, is random, spatially correlated, and second-order stationary. Since we are dealing with only one dimension for the time being, we can replace $\mathbf{x}$ by $x = |\mathbf{x}|$.

and **h** by $h = |\mathbf{h}|$. Its covariance function, in the notation of Chapter 4, is

$$C(h) = \mathrm{E}[\{Z(x) - \mu\}\{Z(x + h) - \mu\}] = \mathrm{E}[Z(x)Z(x + h) - \mu^2], \quad (7.2)$$

where μ is the mean of the process.

The covariance function in the spatial domain has an equivalent in the frequency domain where the variance, instead of being a function of distance (or time), is distributed as a function of frequency, f. This function, denoted by $R(f)$, is the *spectrum*, or *power spectrum*. It is the Fourier transform of the covariance function defined for the interval from positions $-X/2$ to $X/2$, i.e. $-X/2 \leqslant Z(x) \leqslant X/2$:

$$R(f) = \lim_{X \to \infty} \frac{1}{2\pi} \int_{-2X}^{2X} \{1 - (|h|/2X)\} \exp(-\mathrm{j}fh)C(h)\,\mathrm{d}h, \quad (7.3)$$

where j is $\sqrt{-1}$. Provided $C(h)$ approaches 0 as h approaches ∞, the limiting value of $R(f)$ is given by

$$R(f) = \frac{1}{2\pi} \int_{-\infty}^{\infty} \exp(-\mathrm{j}fh)C(h)\,\mathrm{d}h. \quad (7.4)$$

The covariance function is symmetric; it is an 'even' function of h, i.e. $C(h) = C(-h)$; see Chapter 4. As a consequence, the complex term in the integral in equation (7.4) can be replaced by a simple cosine, and $R(f)$ reduces to

$$R(f) = \frac{1}{2\pi} \int_{-\infty}^{\infty} \cos(fh)C(h)\,\mathrm{d}h. \quad (7.5)$$

Just as the spectrum, $R(f)$, is the Fourier transform of the covariance function, the latter, $C(h)$, is the Fourier transform of $R(f)$:

$$C(h) = \frac{1}{2\pi} \int_{-\infty}^{\infty} \cos(fh)R(f)\,\mathrm{d}f. \quad (7.6)$$

In other words, the relation is invertible.

We can equally well transform the autocorrelation function, $\rho(h) = C(h)/C(0)$, to obtain the *normalized spectrum*:

$$r(f) = \frac{1}{2\pi} \int_{-\infty}^{\infty} \cos(fh)\rho(h)\,\mathrm{d}h. \quad (7.7)$$

This relation too is invertible.

7.3.1 Estimating the spectrum

Equations (7.4) and (7.5) above define the spectrum of a real continuous second-order stationary random process in $\mathbb{R}^1$. We want now to estimate the spectrum. As in the example of the gilgai transect, we have data,

$z(x_1), z(x_2), \ldots, z(x_N)$, at regular intervals on a line. The value N is the length of the series, and replaces X to accord with geostatistical convention. From the data we compute

$$\hat{C}(h) = \frac{1}{N - h} \sum_{i=1}^{N-h} \{z(i) - \bar{z}\}\{z(i + h) - \bar{z}\}, \tag{7.8}$$

where the $z(i)$ and $z(i + h)$ are observed values, and $\bar{z}$ is the average of the data in the sequence, and by incrementing h one step at a time we obtain the experimental covariance function. Thus the lag, h, is in units of the sampling interval.

This set of covariances can be transformed to the corresponding experimental spectrum by

$$\hat{R}(f) = \frac{1}{2\pi} \left\{ \hat{C}(0) + 2 \sum_{k=1}^{L-1} \hat{C}(k) \cos(\pi f k) \right\} \tag{7.9}$$

for frequency, f, in the range 0 to $\frac{1}{2}$ cycle. In this equation L is the maximum lag from which the transform is computed and k is the lag.

The quantity L can be regarded as the width of a 'window' through which the covariance is viewed for transformation, and it is for us to choose it. We could set it to the maximum possible from the data. We know from experience that as the lag increases so the experimental covariances become increasingly unreliable, and in Chapter 5 we suggested that the covariance be computed to a lag of no more than about one-fifth of the total length of a series. If we incorporate the uncertainty in estimating $C(h)$ at long lags in equation (7.9) then we shall obtain detail in the computed spectrum that is untrustworthy. In fact, the longer is L, the more detailed is the spectrum and the less reliable is that detail. On the other hand, if we choose too small a value of L then we shall lose detail that might be significant. The window is effectively a smoothing function, and the narrower it is in the spatial domain the more precise are the estimates at the expense of greater bias and loss of detail. So the choice of L is always a compromise.

Some of the fluctuation in the spectrum that arises from choosing a large L can be diminished by changing the 'shape' of the window. The window in equation (7.9) is rectangular: if $|k| \leqslant L$ then $C(k)$ carries weight $1/L$, otherwise its weight is 0:

$$w_R(k) = \begin{cases} 1/L & \text{for } 0 \leqslant |k| \leqslant L, \\ 0 & \text{for } |k| > L, \end{cases} \tag{7.10}$$

and it is shown in Figure 7.3. It is symmetric about the ordinate, and so we show only the positive half. Its Fourier transform is given by

$$W_R(f) = 2L\left(\frac{\sin(2\pi f L)}{2\pi f L}\right) \quad \text{for } -\infty \leqslant f \leqslant \infty. \tag{7.11}$$

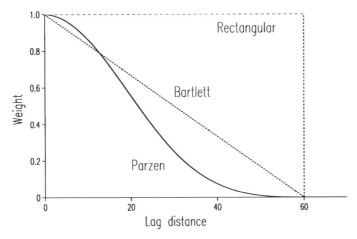

Figure 7.3 Rectangular, Bartlett and Parzen lag windows, with a basal width of 60 sampling intervals.

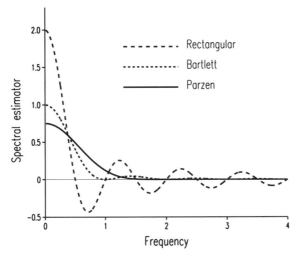

Figure 7.4 Rectangular, Bartlett and Parzen spectral windows. These are the Fourier transforms of the lag windows shown in Figure 7.3.

The transform of the rectangular lag window fluctuates as the frequency increases with a period of $1/L$; the power takes a long while to damp. This is shown in Figure 7.4 in which there are several peaks of decreasing height. In the jargon of spectral analysis, the rectangular window is 'leaky', and it is generally regarded as unsatisfactory. The top corners of the rectangles tend to contribute most of the leakage. This leakage can be diminished substantially by cutting the corners.

Much research has been devoted to finding an optimal shape, 'window carpentry' as Jenkins and Watts (1968) called it. The simplest solution is due to Bartlett (1966) and is known as the Bartlett window. It is defined in the spatial domain as follows:

$$w_B(k) = \begin{cases} 1 - (|k|/L) & \text{for } 0 \leqslant |k| \leqslant L, \\ 0 & \text{for } |k| > L. \end{cases} \tag{7.12}$$

The Bartlett lag window may be envisaged as an isosceles triangle with its peak at its centre and its height decaying linearly to its lower corners where $|k|$ of equation (7.9) equals L. It is shown in Figure 7.3 for $0 \leqslant k \leqslant 0$. Like the spectral window, the lag window is symmetrical about the ordinate, and so again only the positive half is shown. It is incorporated in the transformation equation as

$$\hat{R}(f) = \frac{1}{2\pi} \left\{ \hat{C}(0) + 2 \sum_{k=1}^{L-1} \hat{C}(k) w_B(k) \cos(\pi f k) \right\}. \tag{7.13}$$

The Fourier transform of the Bartlett lag window is

$$W_B(f) = L \left(\frac{\sin(2\pi f L)}{2\pi f L} \right)^2 \qquad \text{for } -\infty \leqslant f \leqslant \infty. \tag{7.14}$$

and this is shown in Figure 7.4. It fluctuates rather less than the rectangular window, but nevertheless is not entirely satisfactory because of its leakage. Two other popular windows are those defined by J. W. Tukey (see Blackman and Tukey, 1958) and Parzen (1961), and these too are referred to by their authors' names. A shortcoming of Tukey's window is that it can return negative estimates of the spectral density, which must be positive. Parzen's window is more reliable. Its definition is

$$w_P(k) = \begin{cases} 1 - 6\left(\dfrac{k}{L}\right)^2 + 6\left(\dfrac{|k|}{L}\right)^3 & \text{for } 0 \leqslant |k| \leqslant L/2, \\ 2\left(1 - \dfrac{|k|}{L}\right)^3 & \text{for } L/2 < |k| \leqslant L, \\ 0 & \text{for } |k| > L, \end{cases} \tag{7.15}$$

and its Fourier transform is

$$W_P(f) = \tfrac{3}{4} L \left(\frac{\sin(\pi f L/2)}{\pi f L/2} \right)^4 \qquad \text{for } -\infty \leqslant f \leqslant \infty. \tag{7.16}$$

In the transformation equation (7.13), for the Parzen window $w_B(k)$ is replaced by $w_P(k)$. Figure 7.4 shows that this transform does not fluctuate, but decays to 0 at a frequency of approximately $2/L$. Although Parzen's window seems the most attractive theoretically, it generally requires a more reliable set of covariances and therefore more data in the first place.

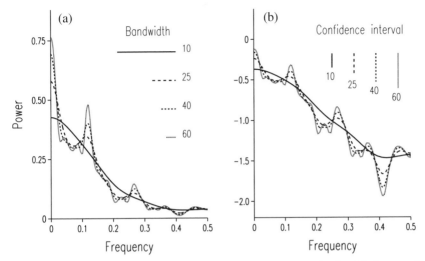

Figure 7.5   Spectrum of log electrical conductivity smoothed with Parzen lag windows of width 10, 25, 40 and 60 sampling intervals on (a) arithmetic scale, and (b) logarithmic scale. Frequency is the reciprocal of sampling interval. The confidence intervals in (b) are for 90%.

To estimate the spectrum from a series of data we compute the experimental covariance function to the maximum lag that is likely to be of interest. This is the initial width, L, of the lag window. We then choose the shape of the window (we recommend the Parzen window as a start), and we compute the spectral density at frequencies between 0 and $\frac{1}{2}$ cycle. We then plot the results and join the points. The steps by which f is incremented need bear no relation to the lag increments, as some authorities claim. In fact, it is better to choose numerous short steps for f so as to produce a smooth figure for the spectrum, which is a continuous function. We then shorten L and repeat the procedure. Figure 7.5 shows results of using this procedure with 100 frequencies.

An alternative method for computing spectra from data is the fast Fourier transform (see Brigham, 1974). Cooley and Tukey (1965) devised an algorithm for its computation in the days when computers were orders of magnitude slower than they are now, and code for it is included in Press *et al.* (1992).

7.3.2   Smoothing characteristics of windows

The windows used to estimate the spectrum are effectively smoothing functions. The estimates have smaller variances than those of the full sample spectrum. To express this quantitatively, we first integrate over the window to obtain a quantity I:

$$I = \int_{-\infty}^{\infty} w^2(k) \, dk. \tag{7.17}$$

So, for example, I for Bartlett's window is the integral from $-L$ to L:

$$I_B = \int_{-L}^{L} \left(1 - \frac{|k|}{L}\right)^2 dk = \tfrac{2}{3}L. \tag{7.18}$$

The equivalent integral of the Parzen window of equation (7.15) is $I_P = L/1.86$.

To distinguish the various spectral estimates we use $R(f)$ to denote the full spectrum, as in equation (7.9), and $\overline{R}$ subscripted with the name of the window and its width for the smoothed estimates. For example, $\overline{R}_{P,L=25}(f)$ means the estimate of $R(f)$ at frequency f smoothed with a Parzen window of width 25 lag intervals. The variance of $R(f)$ is simply $R^2(f)$.

We are interested in the reduction in variance brought about by the smoothing, i.e. the ratio of the variance of the smoothed estimate, $var[\overline{R}(f)]$ to $R^2(f)$. It turns out that this is simply

$$\frac{var[\overline{R}(f)]}{R^2(f)} = \frac{1}{L}\int_{-\infty}^{\infty} w^2(k)\,dk = \frac{I}{L}. \tag{7.19}$$

So, for example, combining this equation with equation (7.18), we find that the variance ratio for a Bartlett window of L/N is $2L/3N = 0.667L/N$. For the Parzen window it is $L/1.86N = 0.538L/N$. Typically L/N is of the order 0.1, and so the variance ratio is around 0.067 for the Bartlett window and 0.054 for the Parzen window—these are big gains in precision.

Bandwidth

As above, each window in the spatial domain has its equivalent in the frequency domain. For a given shape, the wider is the window in the spatial domain the narrower is its transform. Also, because the weight of the lag windows of the same basal width is increasingly concentrated in the order rectangular < Bartlett < Parzen, they behave as if they were increasingly wide in the frequency domain—compare Figures 7.3 and 7.4. It is as though one were viewing the spectrum through a slit of increasing width. The spectral windows do not have strict bounds, however, and it is helpful to have some measure of width for comparison. One approach to this is to define the width of a rectangular window in the frequency domain such that

$$W(f) = \frac{1}{m}, \qquad \text{for } -\tfrac{1}{2}m \leqslant f \leqslant \tfrac{1}{2}m. \tag{7.20}$$

If we denote its bandwidth by b then $b = m$. The variance of the spectral estimator is

$$\text{var}[\overline{R}(f)] = \frac{R^2(f)}{N}\frac{1}{m} = \frac{R^2(f)}{Nb}. \tag{7.21}$$

The bandwidths of the other windows are then defined as those widths that give the same variance as that of the rectangular window,

$$\text{var}[\overline{R}(f)] \approx \frac{R^2(f)}{N}\frac{1}{b} = \frac{R^2(f)}{N}\int_{-\infty}^{\infty} w^2(k)k, \tag{7.22}$$

and so the bandwidth is $b = 1/I$. The bandwidth for the rectangular window is thus $0.5/L$, and that of the Bartlett window is $1.5/L$. The Parzen window's bandwidth is $1.86/L$. Evidently the Bartlett and Parzen windows are substantially wider than the rectangular windows.

7.3.3   Confidence

Confidence intervals for the spectral densities can be determined. We first define a quantity v, which is effectively the degrees of freedom:

$$v = \frac{2N}{\int_{-\infty}^{\infty} w^2(k)\,\mathrm{d}k} = \frac{2N}{I}, \tag{7.23}$$

where N is the total number of observations in the series. Notice that it depends on the width and shape of the window, $w(k)$. The quantity $v\hat{R}(f)/R(f)$ is distributed as χ^2_v, and so

$$\text{Prob}\left[\chi^2_{v,\alpha/2} < \frac{v\hat{R}(f)}{R(f)} \leqslant \chi^2_{v,1-\alpha/2}\right] = 1 - \alpha, \tag{7.24}$$

where Prob stands for the probability and $1 - \alpha$ is the confidence level at which one wants to work. The $100(1 - \frac{1}{2}\alpha)\%$ and $100(\frac{1}{2}\alpha)\%$ confidence limits for $R(f)$ are then

$$\frac{v\hat{R}(f)}{\chi^2_v(1 - \frac{1}{2}\alpha)} \quad \text{and} \quad \frac{v\hat{R}(f)}{\chi^2_v(\frac{1}{2}\alpha)}. \tag{7.25}$$

The integral in equation (7.23) can be worked out for the particular size and shape of window, and, since the length of the sequence, N, is known, v can be determined. The values of χ^2 for v and for $\frac{1}{2}\alpha$ and $1 - \frac{1}{2}\alpha$ can be obtained readily from tables, such as those by Fisher and Yates (1963), or by using a statistical program. This ability to calculate confidence limits gives the spectrum a substantial advantage over the covariance function and variograms.

Table 7.3 Bandwidths and degrees of freedom for the smoothed spectrum of log electrical conductivity at Caragabal.

Lag window	Bartlett window		Parzen window	
	Bandwidth	Deg. freedom	Bandwidth	Deg. freedom
10	0.1500	109.5	0.1860	135.4
25	0.0600	43.8	0.0744	54.2
40	0.0375	27.4	0.0465	33.8
60	0.0250	18.3	0.0310	22.6

7.4   SPECTRAL ANALYSIS OF THE CARAGABAL TRANSECT

The spectrum for the log of electrical conductivity has been computed with a Parzen lag window for four widths: 10, 25, 40 and 60 sampling intervals (Figure 7.5). It is evident that the more covariances are included in the window the more detail there is in the spectrum. One might think there is too much detail with the window set to 60, but with L set to 10 almost all detail has been lost, and only the general decline in power with increasing frequency is evident. Choosing $L = 40$ seems to show the principal features of the spectrum most clearly.

Let us now interpret the spectrum in Figure 7.5. The most prominent feature is the marked decrease in power at the low-frequency end of the spectrum. This corresponds to the spherical and linear components in the variogram. The other striking feature is the peak at around 0.12 cycles. It corresponds to a wavelength of 8.4 sampling intervals or 34 m, which is very close to the wavelength (35 m) of the model fitted to the variogram. Evidently, the spectrum and the variogram are complementary ways of viewing the periodicity and estimating the period.

There is a smaller peak at 0.23 cycles. This is almost certainly a harmonic of the main peak at twice its frequency and may be disregarded. When the spectrum is viewed through a wide window (i.e. computed with the narrowest lag window, $L = 10$ in Figure 7.5) the spectral peak is lost. In this example the bandwidth of the spectral window is much wider than the peak, as Figure 7.5 shows. Therefore, the spectral window must be narrower than the features that one wishes to reveal.

7.4.1   Bandwidths and confidence intervals for Caragabal

In addition to the smoothed spectral estimates, Figure 7.5 shows the bandwidths by the length of the line corresponding to the lag windows 10, 25, 40 and 60. These are calculated for the Parzen windows simply by dividing these widths into 1.86 (Figure 7.5). They are listed in Table 7.3.

The corresponding degrees of freedom, from equation (7.23), are $3N/L$ for the Bartlett window and $3.71N/L$ for the Parzen window, and Table 7.3 also lists their values for the transect.

We can now obtain the confidence limits on the spectral density for any particular frequency. Let us take the Parzen lag window 10. With 365 sampling points this gives $3.71 \times 365/10 = 135.4$ degrees of freedom. If we choose to work at the 90% confidence level, equivalent to $\alpha = 0.1$, then we need χ^2 for $1 - \frac{1}{2}\alpha$ and $\frac{1}{2}\alpha$. These are 109.5 and 163.6, respectively. We now apply equation (7.25). If, for example, we want the confidence limits on our spectral estimate at frequency 0.15, which is $\hat{R}_P(0.15) = 0.2286$, then we calculate

$$c_{\text{lower}} = \frac{135.4}{163.6} \times 0.2286 = 0.189, \qquad c_{\text{upper}} = \frac{135.4}{109.5} \times 0.2286 = 0.282.$$

These could be drawn on Figure 7.5(a), but if you are especially interested in the confidence of spectral estimates it is better to express the intervals on a logarithmic scale. Equation (7.25) becomes

$$\log \hat{R}(f) + \log \frac{\nu}{\chi_\nu^2(1 - \frac{1}{2}\alpha)} \quad \text{and} \quad \log \hat{R}(f) + \log \frac{\nu}{\chi_\nu^2(\frac{1}{2}\alpha)}. \tag{7.26}$$

The interval is constant and symmetric about the logarithm of the estimate. Taking the example above, we compute the logarithm (to base 10) of 0.2286 (which is -0.636) and of the 90% confidence limits. The latter are -0.723 and -0.589, giving a confidence interval of 0.134 in the logarithms. Therefore if the spectrum itself is drawn on a logarithmic scale then the confidence interval can be represented as a single vertical line on the graph.

In Figure 7.5(b) the estimates of Figure 7.5(a) are converted to logarithms, and the results for the 90% confidence intervals are shown by the lengths of the vertical lines. The width of a confidence interval clearly depends on the width of the corresponding lag window. The wider is that window, and the narrower the bandwidth, the wider is the interval.

7.5 FURTHER READING ON SPECTRAL ANALYSIS

The theory of spectral analysis is extensive and complex, and it has numerous applications in many branches of science and engineering. Its principal merits in soil and environmental science are where there is periodicity. It is possible to detect periodicity in variograms and to model it. However, it is often easier to see the periodicity and to estimate it in the spectrum. If periodic variation is suspected from the variogram spectral analysis can be used to confirm that it is present. Oliver *et al.* (1997) used geostatistics and spectral analysis in such a complimentary way.

Two books that deal with spectral analysis at not too advanced a level are by Jenkins and Watts (1968) and by Priestley (1981). The first is intended for engineers, and numerate soil scientists should be able to cope with it. The second, though more mathematical, emphasizes the ideas.

8

Local estimation or prediction: kriging

Most properties of the environment could be measured at any of an infinite number of places, but in practice they are measured at rather few, mainly for reasons of economy. If we wish to know their values elsewhere we must estimate them from the data that we can obtain. The same holds if we want estimates over larger areas for which it has not been possible to measure or observe the properties directly. In Chapter 3 we considered the general problem of estimating values at unsampled places using either a discrete model of spatial variation and classification, or a continuous model with deterministic interpolators. Many statisticians prefer to call the procedure *prediction* to distinguish it from estimating parameters of a distribution. In geostatistics, however, it is almost always called *estimation* for reasons explained by Matheron (1989); we shall use the two terms interchangeably. Estimation is the task for which geostatistics was initially developed, and it is generally called *kriging* after D. G. Krige, a mining engineer in the gold fields of South Africa (see Krige, 1966). The term was coined originally as *krigeage* by P. Carlier, but Matheron (1963) brought it into the English language in recognition of Krige's contribution to improving the precision of estimating concentrations of gold and other metals in ore bodies and recoverable reserves. Although much of the credit for formalizing the technique goes to Matheron and his colleagues at the Paris School of Mines, the mathematics of ordinary kriging had been worked out by A. N. Kolmogorov in the 1930s (Kolmogorov, 1939; 1941; see also Gandin, 1965), by Wold (1938) for time-series analysis, and only a little later by Wiener (1949). You can read a brief history of the subject in Cressie (1990).

8.1 GENERAL CHARACTERISTICS OF KRIGING

Kriging provides a solution to the problem of estimation based on a continuous model of stochastic spatial variation. It makes the best use of existing knowledge by taking account of the way that a property varies in space through the variogram model. In its original formulation a kriged estimate at a place was simply a linear sum or weighted average of the data in its neighbourhood. Since then kriging has been elaborated to tackle increasingly complex problems in mining, petroleum engineering, pollution control and abatement, and public health. The term is now generic, embracing several distinct kinds of kriging, both linear and non-linear. In this chapter we deal with the simpler linear methods, and in Chapter 10 we consider one of the more complex methods, disjunctive kriging. In linear kriging the estimates are weighted linear combinations of the data. The weights are allocated to the sample data within the neighbourhood of the point or block to be estimated in such a way as to minimize the estimation or kriging variance, and the estimates are unbiased. Kriging is optimal in this sense.

8.2 THEORY OF ORDINARY KRIGING

The aim of kriging is to estimate the value of a random variable, Z, at one or more unsampled points or over larger blocks, from more or less sparse sample data on a given support, say $z(\mathbf{x}_1), z(\mathbf{x}_2), \ldots, z(\mathbf{x}_N)$, at $\mathbf{x}_1, \mathbf{x}_2, \ldots, \mathbf{x}_N$. The data may be distributed in one, two or three dimensions, though applications in the environmental sciences are usually two-dimensional.

Ordinary kriging is by far the most common type of kriging in practice, and for this reason we focus on its theory here. It assumes that the mean is unknown. If we consider punctual estimation first, then we estimate Z at a point $\mathbf{x}_0$ by $\hat{Z}(\mathbf{x}_0)$, with the same support as the data, by

$$\hat{Z}(\mathbf{x}_0) = \sum_{i=1}^{N} \lambda_i z(\mathbf{x}_i), \tag{8.1}$$

a weighted average of the data, where λ_i are the weights. To ensure that the estimate is unbiased, the weights are made to sum to 1,

$$\sum_{i=1}^{N} \lambda_i = 1,$$

and the expected error is $\mathrm{E}[\hat{Z}(\mathbf{x}_0) - Z(\mathbf{x}_0)] = 0$. The estimation variance is

$$\mathrm{var}[\hat{Z}(\mathbf{x}_0)] = \mathrm{E}[\{\hat{Z}(\mathbf{x}_0) - Z(\mathbf{x}_0)\}^2]$$

$$= 2\sum_{i=1}^{N} \lambda_i \gamma(\mathbf{x}_i, \mathbf{x}_0) - \sum_{i=1}^{N}\sum_{i=j}^{N} \lambda_i \lambda_j \gamma(\mathbf{x}_i, \mathbf{x}_j), \tag{8.2}$$

where $y(\mathbf{x}_i, \mathbf{x}_j)$ is the semivariance of Z between the data points $\mathbf{x}_i$ and $\mathbf{x}_j$, and $y(\mathbf{x}_i, \mathbf{x}_0)$ is the semivariance between the ith data point and the target point $\mathbf{x}_0$.

In the more general case we may wish to estimate Z in a block B, which may be a line, an area or a volume depending on whether it is in one, two or three spatial dimensions. The kriged estimate in B is still a simple weighted average of the data,

$$\hat{Z}(B) = \sum_{i=1}^{N} \lambda_i z(\mathbf{x}_i), \tag{8.3}$$

but with $\mathbf{x}_0$ of equation (8.1) replaced by B. Its variance is

$$\mathrm{var}[\hat{Z}(B)] = \mathrm{E}[\{\hat{Z}(B) - Z(B)\}^2]$$

$$= 2 \sum_{i=1}^{N} \lambda_i \overline{y}(\mathbf{x}_i, B) - \sum_{i=1}^{N} \sum_{i=j}^{N} \lambda_i \lambda_j y(\mathbf{x}_i, \mathbf{x}_j) - \overline{y}(B, B). \tag{8.4}$$

The quantity $\overline{y}(\mathbf{x}_i, B)$ is the average semivariance between the ith sampling point and the block B and is the integral

$$\overline{y}(\mathbf{x}_i, B) = \frac{1}{|B|} \int_B y(\mathbf{x}_i, \mathbf{x}) \, \mathrm{d}\mathbf{x}, \tag{8.5}$$

where $y(\mathbf{x}_i, \mathbf{x})$ denotes the semivariance between the sampling point $\mathbf{x}_i$ and a point $\mathbf{x}$ describing the block, Figure 8.1(a). The third term on the right-hand side of equation (8.4) is the double integral

$$\overline{y}(B, B) = \frac{1}{|B|^2} \int_B \int_B y(\mathbf{x}, \mathbf{x}') \, \mathrm{d}\mathbf{x}\mathrm{d}\mathbf{x}', \tag{8.6}$$

(a)   (b)

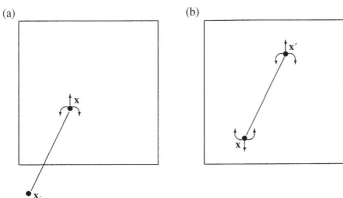

Figure 8.1   Integration of the variogram: (a) between a sampling point and a block; (b) within a block.

where $y(\mathbf{x}, \mathbf{x}')$ is the semivariance between two points $\mathbf{x}$ and $\mathbf{x}'$ that sweep independently over B, Figure 8.1(b). It is the within-block variance. In punctual kriging $\bar{y}(B, B)$ becomes $y(\mathbf{x}_0, \mathbf{x}_0) = 0$, which is why there is one less term in equation (8.2).

For each kriged estimate there is an associated kriging variance, which we can denote by $\sigma^2(\mathbf{x}_0)$ and $\sigma^2(B)$ for the point and block, respectively, and which are defined in equations (8.2) and (8.4). The next step in kriging is to find the weights that minimize these variances, subject to the constraint that they sum to 1. This is achieved using the method of Lagrange multipliers.

We define an auxiliary function $f(\lambda_i, \psi)$ that contains the variance we wish to minimize plus a term containing a Lagrange multiplier, ψ. For punctual kriging it is

$$(\lambda_i, \psi) = \operatorname{var}[\hat{Z}(\mathbf{x}_0) - Z(\mathbf{x}_0)] - 2\psi \left\{ \sum_{i=1}^{N} \lambda_i - 1 \right\}. \tag{8.7}$$

We then set the partial derivatives of the function with respect to the weights to 0:

$$\frac{\partial f(\lambda_i, \psi)}{\partial \lambda_i} = 0,$$
$$\frac{\partial f(\lambda_i, \psi)}{\partial \psi} = 0, \tag{8.8}$$

for $i = 1, 2, \ldots, N$. This leads to a set of $N+1$ equations in $N+1$ unknowns:

$$\sum_{i=1}^{N} \lambda_i y(\mathbf{x}_i, \mathbf{x}_j) + \psi(\mathbf{x}_0) = y(\mathbf{x}_j, \mathbf{x}_0), \qquad \text{for all } j,$$
$$\sum_{i=1}^{N} \lambda_i = 1. \tag{8.9}$$

This is the ordinary kriging system for points. Its solution provides the weights, λ_i, which are entered into equation (8.1), and from which the estimation variance (prediction variance or specifically kriging variance) can be obtained as

$$\sigma^2(\mathbf{x}_0) = \sum_{i=1}^{N} \lambda_i y(\mathbf{x}_i, \mathbf{x}_0) + \psi(\mathbf{x}_0). \tag{8.10}$$

If a target point, $\mathbf{x}_0$, happens to be one of the data points, say $\mathbf{x}_j$, then $\sigma^2(\mathbf{x}_0)$ is minimized when $\lambda(\mathbf{x}_j) = 1$ and all of the other weights are 0. In fact, $\sigma^2(\mathbf{x}_0) = 0$, and by inserting the weights into equation (8.1) we obtain the recorded value, $z(\mathbf{x}_j)$, as our estimate of $z(\mathbf{x}_0)$. Punctual kriging is thus an exact interpolator.

The equivalent kriging system for blocks is

$$\sum_{i=1}^{N} \lambda_i \gamma(\mathbf{x}_i, \mathbf{x}_j) + \psi(B) = \overline{\gamma}(\mathbf{x}_j, B), \qquad \text{for all } j,$$

$$\sum_{i=1}^{N} \lambda_i = 1,$$

(8.11)

with the associated variance obtained as

$$\sigma^2(B) = \sum_{i=1}^{N} \lambda_i \overline{\gamma}(\mathbf{x}_i, B) + \psi(B) - \overline{\gamma}(B, B).$$

(8.12)

The kriging equations can be represented in matrix form. For punctual kriging they are

$$\mathbf{A}\boldsymbol{\lambda} = \mathbf{b},$$

(8.13)

where

$$\mathbf{A} = \begin{bmatrix} \gamma(\mathbf{x}_1, \mathbf{x}_1) & \gamma(\mathbf{x}_1, \mathbf{x}_2) & \cdots & \gamma(\mathbf{x}_1, \mathbf{x}_N) & 1 \\ \gamma(\mathbf{x}_2, \mathbf{x}_1) & \gamma(\mathbf{x}_2, \mathbf{x}_2) & \cdots & \gamma(\mathbf{x}_2, \mathbf{x}_N) & 1 \\ \vdots & \vdots & \cdots & \vdots & \vdots \\ \gamma(\mathbf{x}_N, \mathbf{x}_1) & \gamma(\mathbf{x}_N, \mathbf{x}_2) & \cdots & \gamma(\mathbf{x}_N, \mathbf{x}_N) & 1 \\ 1 & 1 & \cdots & 1 & 0 \end{bmatrix},$$

$$\boldsymbol{\lambda} = \begin{bmatrix} \lambda_1 \\ \lambda_2 \\ \vdots \\ \lambda_N \\ \psi(\mathbf{x}_0) \end{bmatrix},$$

and

$$\mathbf{b} = \begin{bmatrix} \gamma(\mathbf{x}_1, \mathbf{x}_0) \\ \gamma(\mathbf{x}_2, \mathbf{x}_0) \\ \vdots \\ \gamma(\mathbf{x}_N, \mathbf{x}_0) \\ 1 \end{bmatrix}.$$

Matrix $\mathbf{A}$ is inverted, and the weights and the Lagrange multiplier are obtained as

$$\boldsymbol{\lambda} = \mathbf{A}^{-1}\mathbf{b}.$$

(8.14)

The kriging variance is given by

$$\hat{\sigma}^2(\mathbf{x}_0) = \mathbf{b}^{\mathrm{T}}\boldsymbol{\lambda}. \tag{8.15}$$

For block kriging the only differences are that

$$\mathbf{b} = \begin{bmatrix} \overline{\gamma}(\mathbf{x}_1, B) \\ \overline{\gamma}(\mathbf{x}_2, B) \\ \vdots \\ \overline{\gamma}(\mathbf{x}_N, B) \\ 1 \end{bmatrix}$$

and

$$\hat{\sigma}^2(B) = \mathbf{b}^{\mathrm{T}}\boldsymbol{\lambda} - \overline{\gamma}(B, B). \tag{8.16}$$

8.3   WEIGHTS

When the kriging equations are solved to obtain the weights, λ_i, in general the only large weights are those of the points near to the point or block to be kriged. The nearest four or five might contribute 80% of the total weight, and the next nearest ten almost all of the remainder. The weights also depend on the configuration of the sampling. We can summarize the factors affecting the weights as follows.

1. Near points carry more weight than more distant ones. Their relative proportions depend on the positions of the sampling points and on the variogram: the larger is the nugget variance, the smaller are the weights of the points that are nearest to target point or block.

2. The relative weights of points also depend on the block size: as the block size increases the weights of the nearest points decrease and those of the more distant points increase (Figure 8.9) until the weights become nearly equal.

3. Clustered points carry less weight individually than isolated ones at the same distance (Figure 8.12).

4. Data points can be screened by ones lying between them and the target (Figure 8.12).

These effects are all intuitively desirable, and the first shows that kriging is local. They will become evident in the examples below. They also have practical implications. The most important for present purposes is that because only the nearest few data points to the target carry significant weight, matrix $\mathbf{A}$ in the kriging system need never be large and its inversion will be swift. We can replace N in equations (8.9) and (8.11) by some much smaller number, say $n \ll N$. We shall reiterate this below after the examples in which we set n to 16.

8.4 EXAMPLES

This section shows the effects of a changing variogram, target point and sampling intensity on the weights in a way analogous to the kriging exercises in GSLIB (Deutsch and Journel, 1992). It uses the data on pH from Broom's Barn Farm for the purpose. We have chosen pH because it is easy to appreciate changes in the estimated values and because we can start with a simple isotropic exponential model without nugget (Figure 8.2), which is the best-fitting model:

$$\gamma(h) = c\{1 - \exp(-h/r)\}, \tag{8.17}$$

with $c = 0.382$ and $r = 90.53$ m, i.e. an effective range ($a' = 3r$) of approximately 272 m (Table 8.1). We have also selected $n = 16$ points on a 4×4 lattice from the full set of data (Figure 8.3). There are also three separate target points, one at the centre of the lattice, Figure 8.3(a), one off-centre, Figure 8.3(b), and a third coinciding with one of the sampling points, Figure 8.11(c). Using equation (8.9) and the 16 points, we estimated the values at the target points as follows.

8.4.1 Kriging at the centre of the lattice

Changing the ratio of nugget: sill variance

Figure 8.4(a) shows the weights derived using the best-fitting model to pH (exponential (N1), Table 8.1; and exponential (R2), Table 8.2). The weights of the four points nearest to the target point are large and positive, and their sum exceeds 1. To ensure unbiasedness the sum of all the weights must be 1, and hence the weights of the outer points are negative. In this

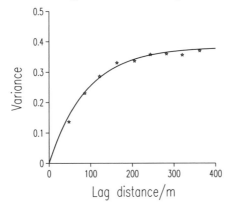

Figure 8.2 Variogram of pH at Broom's Barn Farm. The points are the experimental semivariances, and the solid line is the best-fitting exponential model; the parameters of which are given in the text.

Table 8.1 Model parameters with changing ratio of nugget:sill variance and fixed distance parameter, $r = 90.53$ m, equivalent to an effective range of 271.59 m.

Model	c_0	c
Exponential (N1)	0	0.3820
Exponential (N2)	0.1	0.2820
Exponential (N3)	0.3	0.0820
Exponential (N4)	0.382	0
(Pure nugget)	1	0

(a)   (b)

```
8.0      8.0      8.0      8.0          8.0      8.0      8.0      8.0
 •        •        •        •            •        •        •        •


7.9      8.0      7.8      7.8          7.9      8.0      7.8      7.8
 •        •        •        •            •        •        •        •
                 ?                                       ?
              ⊗                                       ⊗


 •        •        •        •            •        •        •        •
6.9      6.2      6.2      6.2          6.9      6.2      6.2      6.2


 •        •        •        •            •        •        •        •
7.0      6.0      5.8      6.0          7.0      6.0      5.8      6.0
```

Figure 8.3 The grid of 16 sample values selected from Broom's Barn Farm with the pH values given for each sampling location. The point to be estimated is located: (a) centrally; (b) off-centre.

case the outer points are close to 0 and so have little influence on the estimate.

We now change the variogram by introducing a nugget variance, $c_0 = 0.1$ (the model parameters are those of exponential (N2) in Table 8.1). The resulting weights are shown in Figure 8.4(b): those of the inner four points have decreased somewhat, whilst those of the outer points have increased and are now all positive. The weights of the corner points of the lattice are the smallest because they are the furthest from x_0.

Figure 8.4(c) shows the weights obtained by increasing the proportion of nugget more substantially so that it dominates (the model parameters are those of exponential (N3) in Table 8.1). The weights of the inner points have decreased considerably, and those of the outer ones have increased correspondingly.

For a pure nugget variogram, with parameters exponential (N4) in Table

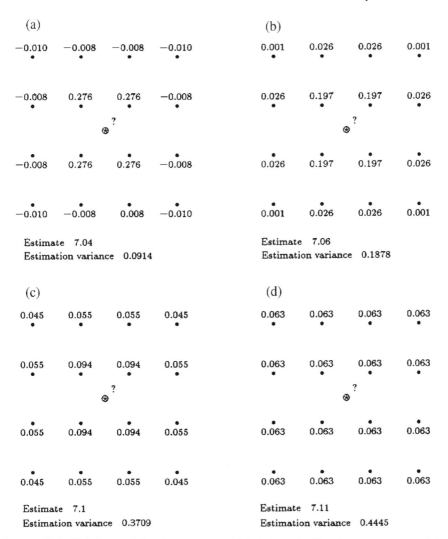

Figure 8.4 Kriging weights from punctual kriging of pH using an exponential function with the distance parameter $r = 90.53$ m, and changing the nugget:sill variance: (a) $c_0 = 0$, $c = 0.382$; (b) $c_0 = 0.1$, $c = 0.282$; (c) $c_0 = 0.3$, $c = 0.082$; (d) $c_0 = 0.382$, $c = 0$.

8.1, the weights are all the same, Figure 8.4(d). The result is the same as if we had sampled at random in classical estimation; the kriging variance is the variance of the process, c_0, plus the variance of the mean, given by $\psi(\mathbf{x}_0)$. The solution of equation (8.15) is

$$c_0 + \psi(\mathbf{x}_0) = 0.382 + 0.0625 = 0.4445. \qquad (8.18)$$

Figure 8.5(a) summarizes the shapes of the exponential variogram mod-

Table 8.2   Model parameters with changing distance parameter, r, for exponential model.

Model	c_0	c	r (m)	Effective range (m)
Exponential (R1)	0	0.382	133.33	400.00
Exponential (R2)	0	0.382	90.53	272.00
Exponential (R3)	0	0.382	26.67	80.00
Exponential (R4)	0	0.382	6.67	20.00

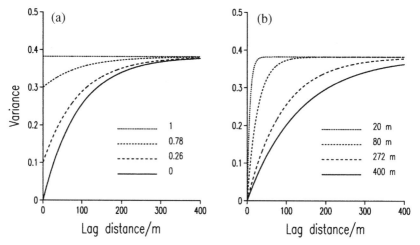

Figure 8.5   (a) Exponential variograms used to obtain the weights in Figure 8.4 with the distance parameter $r = 90.53$ m and constant sill, $c_0 + c = 0.382$, and nugget : sill variance ratio 0 (no nugget), 0.26, 0.78 and 1 (pure nugget). (b) Exponential variograms with $c_0 = 0$ and $c = 0.382$ and four effective ranges ($a' = 3r$), $a' = 400$ m, 272 m, 80 m and 20 m.

els that resulted from changing the ratio of nugget : sill variance and keeping the distance parameter constant.

The estimated values for pH and estimation variance are given for each of the above examples (Figure 8.4). The estimated value changes each time: we can assume that 7.04 is the optimal estimate because it was derived from the best-fitting model. The average pH of the 16 values is 7.11, which is also the estimate returned with a pure nugget variogram and for which the estimation variance is the largest. The estimation variance increases as the nugget variance increases, as we should expect, because the greater the variance that remains unresolved the greater is the uncertainty in the estimate. The estimates and estimation variances illustrate two points:

(i) A nugget variance increases the estimation variance, and for punctual kriging it sets a lower limit to the estimation variance, see Figures 8.17 and 10.8(b).

(ii) It is important to fit the model correctly to the sample semivariances because of the effect of the model on both the estimates and estimation variances.

Although the estimation variances are smaller if the nugget variance is smaller, the model must reflect the nugget realistically. If it does not then the estimates could be judged to be more or less reliable than they really are.

Changing the range or sampling intensity

We now explore the effect of decreasing the range of spatial dependence, or, what amounts to the same thing, decreasing the sampling density. The nugget variance and the sill of the spatially dependent component, c, were kept constant and we changed the distance parameter, as shown in Table 8.2. Figure 8.5(b) shows the effect on the shape of the exponential variogram, and Figure 8.6(a) shows the weights for exponential (R1), Table 8.2, where the effective range of dependence is 400 m. The weights of the inner four points are the largest, and the outer ones contribute little or nothing. If we compare this with Figure 8.6(b) for the best-fitting exponential model (R2), Table 8.2, it is clear that they are similar. As the effective range lengthens, however, the inner points gain weight in accordance with the increase in spatial continuity in the variation. If we reduce the effective range substantially to 80 m (exponential (R3), Table 8.2) the weights of the inner points decrease and those of the outer ones increase, Figure 8.6(c). When we reduce the effective range to half of the sampling interval, i.e. 20 m (exponential (R4), Table 8.2), the variogram is in essence pure nugget. Figure 8.6(d) shows the weights which are now small for all of the points, though they are not all the same: the inner ones are somewhat larger than the outer ones, because with the exponential model the distance parameter does not disappear completely. Nevertheless, the estimate is the mean of the data as in the previous example, Figure 8.4(d), but the estimation variance is a little less because of the effect of the differences in the weights.

Since changing the distance parameter of a spherical model has a different effect, we include the results of changing the range of the best-fitting spherical function to the 16 points. The spherical function is given by

$$y(h) = c_0 + c\left\{\frac{3h}{2a} - \frac{1}{2}\left(\frac{h}{a}\right)^3\right\}, \tag{8.19}$$

with the parameter values $c_0 = 0.0309, c = 0.3211$ and $a = 203.2$ m for the best-fitting spherical function.

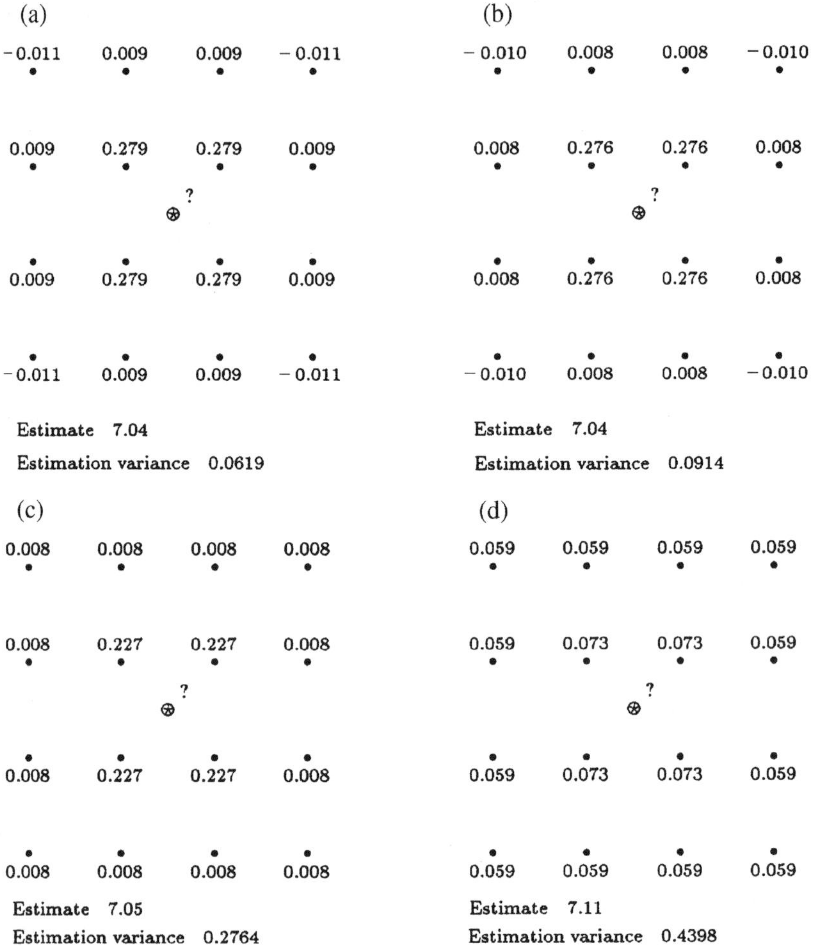

(a)

-0.011	0.009	0.009	-0.011
0.009	0.279	0.279	0.009
0.009	0.279	0.279	0.009
-0.011	0.009	0.009	-0.011

Estimate 7.04

Estimation variance 0.0619

(b)

-0.010	0.008	0.008	-0.010
0.008	0.276	0.276	0.008
0.008	0.276	0.276	0.008
-0.010	0.008	0.008	-0.010

Estimate 7.04

Estimation variance 0.0914

(c)

0.008	0.008	0.008	0.008
0.008	0.227	0.227	0.008
0.008	0.227	0.227	0.008
0.008	0.008	0.008	0.008

Estimate 7.05

Estimation variance 0.2764

(d)

0.059	0.059	0.059	0.059
0.059	0.073	0.073	0.059
0.059	0.073	0.073	0.059
0.059	0.059	0.059	0.059

Estimate 7.11

Estimation variance 0.4398

Figure 8.6 Kriging weights from punctual kriging of pH using an exponential function with $c_0 = 0.0$ and $c = 0.382$, and changing the effective range ($a' = 3r$): (a) $a' = 400$; (b) $a' = 272$; (c) $a' = 80.0$; (d) $a' = 20.0$.

There are several interesting differences between the results of these models. Figure 8.7(a) shows the best-fitting spherical and exponential models fitted to pH, and Figure 8.7(b) shows the effect of changing the range on the shape of the spherical model.

To compare the weights with those for the exponential model, we started with spherical (A1), Table 8.3, with a range of 400 m. The weights of the inner points are less, and those of the outer ones more, than those for the exponential model, Figure 8.8(a). This is because the spherical model has a small nugget variance, whereas the exponential had none, and there is a difference in the curvature of these two models (Figure 8.7(a)). Fig-

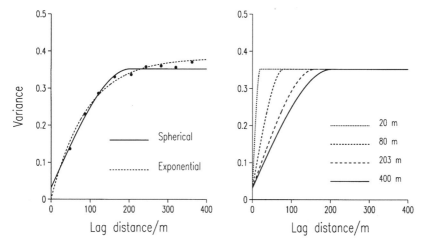

Figure 8.7 (a) The best-fitting spherical and exponential functions for the variogram at Broom' Barn Farm. (b) Spherical variograms used to obtain the weights in Figure 8.8 with $c_0 = 0.031$, $c = 0.321$, and ranges $a = 400$ m, 203.2 m (the same as in (a)), 80 m and 20 m.

Table 8.3 Model parameters with the range changing for spherical model.

Model	c_0	c	Range (m)
Spherical (A1)	0.0309	0.3211	400.0
Spherical (A2)	0.0309	0.3211	203.2
Spherical (A3)	0.0309	0.3211	80.0
Spherical (A4)	0.0309	0.3211	20.0

ure 8.8(b) shows the weights obtained when using the best-fitting spherical function (spherical (A2), Table 8.3); the inner weights are larger and the outer ones slightly less. It is a reversal of the effect with the exponential model. When the range is reduced to 80 m (spherical (A3), Table 8.3) the inner weights, see Figure 8.8(c), are much larger than for the equivalent exponential model, see Figure 8.6(c), again because of the effect of the model's curvature. Finally, when the range is 20 m the weights are all the same, see Figure 8.8(d), and the observed effect is the same as that for the pure nugget variogram. In this situation all of the variation occurs within the sampling interval. It illustrates clearly that if the distance over which most of the variation occurs is less than the sampling interval, then the simple formula for random sampling gives the best estimate for an unsampled point, which is the mean of the data. It also shows the importance of sampling sufficiently densely to estimate the variogram at the spatial scale of the investigation.

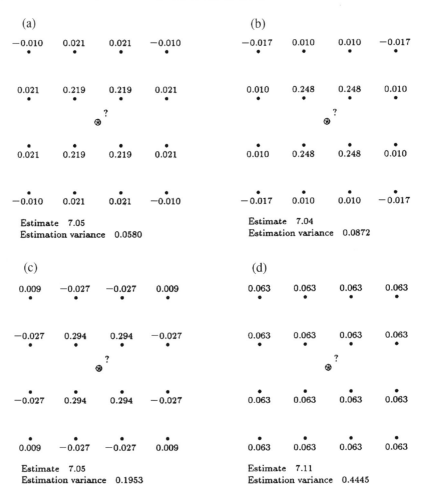

Figure 8.8 Kriging weights from punctual kriging of pH using a spherical function with $c_0 = 0.031$, $c = 0.321$, and changing the distance parameter (range): (a) $a = 400$ m; (b) $a = 203.2$ m; (c) $a = 80$ m; (d) $a = 20$ m.

In summary Figure 8.8(a)–(c) shows that as the distance parameter decreases the weights of the inner points increase and those of the outer ones decrease. Apart from Figure 8.8(a), the estimates are sensibly the same as those for the exponential model, but the estimation variances for the spherical model are smaller in every case.

Kriging over a block

The results for block kriging over a centrally located 80 m × 80 m block using the parameters of the best-fitting exponential model (N1), Table 8.1, are shown in Figure 8.9(a), and those using exponential (N2) in Figure

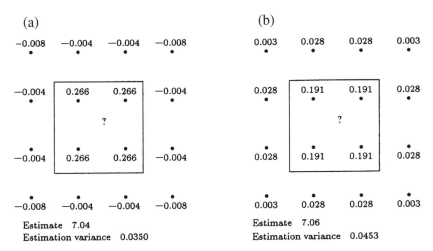

Figure 8.9   Kriging weights from block kriging of pH over a centrally located block of 80 m × 80 m: (a) using the best-fitting exponential model with $c_0 = 0$, $c = 0.382$ and $r = 90.53$ m; (b) $c_0 = 0.1$, $c = 0.282$ and $r = 90.53$ m.

8.9(b). A comparison of the weights in Figures 8.4(a) and 8.9(a) shows that by increasing the block size the inner weights decrease and the outer ones increase. The differences between the two figures are small, but the estimation variance for block kriging with this model is only a little more than a third of that for punctual kriging. With a modest nugget variance, exponential (N2), the relative decrease in the inner weights for block kriging, Figure 8.9(b), is somewhat less than in Figure 8.4(b), but the decrease in the kriging variance over that of punctual kriging is even more marked; it is now less than a quarter. Nevertheless, the estimated values are the same in each case.

This comparison shows two effects of the nugget variance:

1. The nugget variance sets a lower limit to the punctual kriging variance.

2. The nugget variance disappears from the block kriging variance, see equations (8.4) and (8.12). Therefore, the larger is the proportion of the nugget variance, which is taken out of consideration, the smaller is the block kriging variance and the greater is the difference between it and the punctual kriging variance.

It also raises an important issue of confidence. When practitioners fit models to variograms, whether by eye or by minimizing some function of the residuals, they project their models towards the ordinate with the least change in curvature. They know nothing about the shape of the variogram at distances less than the shortest lag interval, and the practice may be regarded as prudent. The intercept gives them a nugget variance which is almost certainly larger than any error of measurement or short-range spatial component. When the model is used for punctual kriging the errors will, therefore, tend to be on the large side; the estimates are conservative. However, when the same model is used for block kriging, if the nugget variance is exaggerated then the kriging variance will be too small for the reasons given above, and the practitioner will obtain a false sense of confidence.

The effect of anisotropy

We examine the effect of geometric anisotropy on the weights by punctual kriging with the following exponential model:

$$\gamma(h, \theta) = c\{1 - \exp(-h/\Omega)\}, \tag{8.20}$$

where Ω is given by

$$\Omega = \sqrt{A^2 \cos^2(\theta - \phi) + B^2 \sin^2(\theta - \phi)}, \tag{8.21}$$

in which $A = 271.6, B = 90.5$ and $\phi = \pi/2 = 0.7854$ rad or $45°$. The angle ϕ is the direction of maximum continuity, i.e. largest effective range, as in Figure 6.11. The ratio A/B is the anisotropy ratio, which is $3 = 271.6/90.5$. Figure 8.10(a) shows the weights for the isotropic variogram and Figure 8.10(b) those for the anisotropic function. The largest weights are at the points adjacent to the target point along the $45°$ diagonal.

There is a marked decrease in the weights of the adjacent points at right angles. The changes in the weights of the outer points are far less marked. If we change ϕ to 0.2618 rad, or $15°$ ($75°$ in geographical notation), then Figure 8.10(c) ensues; the distribution of the weights has changed. The increase in the weights of the nearest points in less dramatic, but the outer weights close to the ($15°$) line have increased substantially to 0.153.

8.4.2   Kriging off-centre in the lattice and at a sampling point

Let us now use the exponential models, one with no nugget (N1) and the other with a nugget variance of 0.1 (N2), Table 8.1, to estimate the value at a target point that is off-centre but on a diagonal of the grid (Figure 8.11). In both Figure 8.11(a) and 8.11(b) the point closest to the target has the largest weight, and the point diagonally opposite has the smallest weight of the four inner points. The other two inner points have the same weights because they are equidistant from the target. The weights of the outer points now show the effect of screening. Figure 8.11(a) shows that the unscreened outer points have positive weights, whereas those that are screened are negative.

The weights in Figure 8.11(c) were obtained by solving the kriging system for punctual kriging at the target point coinciding with the sampling

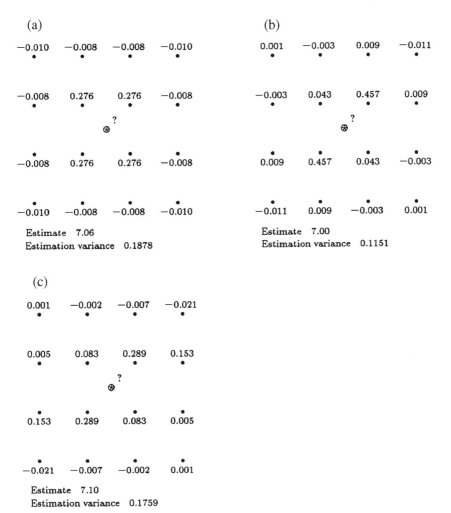

Figure 8.10 Kriging weights from punctual kriging of pH: (a) using the best-fitting exponential model with $c_0 = 0$, $c = 0.382$ and $r = 90.53$ m; (b) using an anisotropic exponential model with the direction of maximum variation $\pi/2$ radians and an anisotropy ratio of 3; (c) with the direction of maximum variation 1.309 rad.

location indicated. They are 1 at the sampling point and 0 elsewhere, which is what we should expect from theory. The estimate is the sample value, and the estimation variance is 0, so illustrating that ordinary punctual kriging is an exact interpolator. The weights in Figure 8.11(d) were obtained using the same model and kriging over a 80 m × 80 m block centred at the same sampling point. The weight at the sampling location is an order of magnitude larger than those of the surrounding nearest

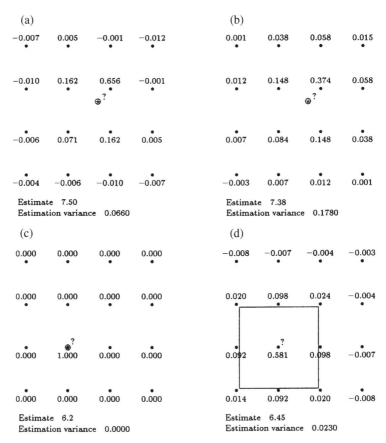

Figure 8.11   Kriging weights of pH using the exponential function: punctual kriging with the point to be estimated off-centre and the model (a) $c_0 = 0$, $c = 0.382$, $r = 90.53$ m, and (b) $c_0 = 0.1$, $c = 0.282$, $r = 90.53$ m; (c) with the point to be estimated at a sampling location with $c_0 = 0$, $c = 0.382$ and $r = 90.53$ m; and (d) block kriging with an 80 m × 80 m block centred at a sampling point with $c_0 = 0$, $c = 0.382$ and $r = 90.53$ m.

points, while those of the outer edges are negative. The estimate is substantially different from the measured value at the centre of the block and shows the smoothing effect of block kriging. The estimation variance is also very small, but not zero.

8.4.3   Kriging from irregularly spaced data

Figure 8.12 shows an irregular configuration of nine sampling points plus a target point; the nine are a selection from the 16 values used previously, but some of the locations were changed. The weights in Figure 8.12(a) were obtained using the best-fitting exponential model (N1), Table 8.1, and

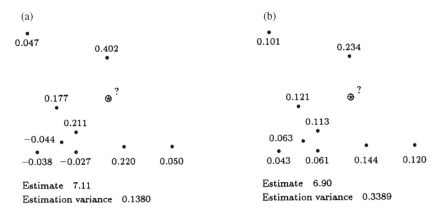

Figure 8.12 Kriging weights from punctual kriging of pH with an exponential model and irregularly scattered sampling points: (a) $c_0 = 0$, $c = 0.382$ and $r = 90.53$ m; (b) $c_0 = 0.2$, $c = 0.182$ and $r = 90.53$ m;

punctual kriging. Those in Figure 8.12(b) were derived with exponential (N2), Table 8.1. The two diagrams show more clearly than those for the grid the effect of the data configuration on the weights. Points that are clustered carry less weight relative to isolated ones. The point to the north of the target carries almost twice the weight of the next most important point because it is far from any other point. The points that are screened by others have negative weights.

8.5 NEIGHBOURHOOD

The notion of the neighbourhood embodies the local nature of kriging, and it confers advantages on the method, as follows.

1. Only the nearest few points to the target point or block carry significant weight, therefore the kriging system need never be large and inverting matrix **A** will be swift. We can replace N in equations (8.9) and (8.11) by a much smaller $n \ll N$. This might not matter when kriging only one point or block, but for mapping in which many estimates are needed it can make a big difference because the time required to invert a matrix is approximately proportional to the cube of its order. It also avoids instability that can arise with large matrices.

2. If only the points near to the target carry significant weight then the variogram need be estimated and modelled well only at short lag distances, and in fact this is usually where the variogram is best estimated. The widening of the confidence intervals on the experimental variogram is somewhat less serious than it might appear

from Chapter 5. This adds to the desirability of giving most weight to the experimental semivariances at the short lags when modelling the variogram.

3. The local nature of ordinary kriging means that what happens over large distances is of little consequence for the estimates. We can accept the notion of quasi-stationarity, i.e. local stationarity (Chapter 4), compute the variogram over only short distances, and apply it without taking account of long-range fluctuations in $E[Z(\mathbf{x})]$. The assumptions underpinning the method are not violated. It is perhaps this feature that has made ordinary kriging the 'workhorse' of geostatistics.

There are no strict rules for defining the neighbourhood, but we suggest some guidelines:

1. If the variogram is bounded and has a small nugget variance and the data are dense then the radius of the neighbourhood can be set close to the range or effective range. Any data beyond the range will have negligible weights.

2. If data are sparse, however, points beyond the range from the target might carry sufficient weight to be important, and the neighbourhood should be such as to include them.

3. If the nugget variance is large, then again distant points are likely to carry significant weight.

4. As an alternative, the user may choose the nearest n data points, and effectively let this number limit the neighbourhood. If the sampling configuration is irregular then the size of the neighbourhood will vary more or less as the target point is moved. We have found that a maximum of $n \approx 20$ is usually enough.

5. If you set a maximum radius for the neighbourhood then you may also need to set a minimum for n, especially to cater for targets near the boundary of a region. A value of $n \approx 7$ is likely to be satisfactory.

6. Where the scatter is very uneven, good practice is to divide the space around the target point into octants and take the nearest two points in each. Several kriging programs do this as a matter of course.

We recommend that when you start to analyse new data you examine what happens to the kriging weights as you change the neighbourhood. This is especially important in mapping where you move the neighbourhood. In these circumstances the most distant points should have zero weight so that the estimated surface appears seamless; see Laslett *et al.* (1987) for an illustrated discussion.

8.6 ORDINARY KRIGING FOR MAPPING

Kriging was developed in mining originally to estimate the amounts of metal in blocks of rock, and it is still used in this way. In these circumstances every block of rock is potentially of interest, and its metal content will be estimated. The miner may then decide whether the rock contains sufficient metal to be mined and sent for processing.

Environmental scientists, and pedologists in particular, have used kriging in a rather different way, namely optimal interpolation at many places for mapping. The earliest examples are those by Burgess and Webster (1980a; 1980b), and Burgess *et al.* (1981), who used ordinary kriging. There have been many since, for example Mulla (1997), Frogbrook (1999) and Frogbrook *et al.* (1999) in precision agriculture.

To map a variable the values are kriged at the nodes of a fine grid. Isarithms are then threaded through this grid, and there are now many programs and packages, such as Surfer (Golden Software, 1997) and Gsharp, and geographical information systems, such as Arc/Info, that will do this with excellent graphics. Computing the isarithms involves another interpolation which is rarely optimal in the kriging sense, but if the kriged grid is fine enough this lack of optimality is immaterial. In most instances kriging at intervals of 2 mm on the finished map will be adequate.

The kriging variances and their square roots, the kriging errors, can be mapped similarly, and these maps give an idea of the reliability of the maps of estimates.

Creating a grid of kriged values to make a map can involve heavy computation. In principle all the estimates and their variances could be found from a single inversion of matrix A in equation (8.13) that contains all of the semivariances between the sampling sites. As above, however, this is unwise or even impossible when the matrices are large. In practice, therefore, one enters into A only the semivariances for some n data points, i.e. within the neighbourhood, near each grid node. This keeps the matrix small, but increases the number of inversions needed. Inversion can be accelerated by working with the covariances instead of the semivariances because in the usual method of matrix inversion the largest element in each row, which serves as a pivot, is always in the diagonal of the covariance matrix and need not be sought.

For variables that are second-order stationary all the formulae for finding the weights from the variogram also apply to the covariance function with only changes of sign. For variables that are intrinsic only, the technique can still be used by taking some arbitrary large value for the covariance at $h = 0$.

Other economies can be made depending on the location of the sampling points. If they are irregularly scattered then the same few data will often be used to estimate $Z(x)$ at several grid nodes within a small area.

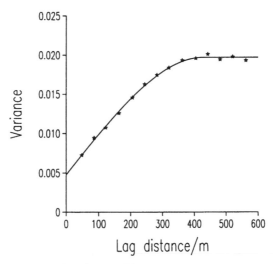

Figure 8.13   Variogram of exchangeable potassium at Broom's Barn Farm transformed to common logarithms. The points are the experimental semivariances, and the solid line is the best-fitting spherical model, the parameters of which are given in the text.

Furthermore, the finer the interpolation grid the more nodes can be interpolated from the same observations. Matrix **A** remains the same and needs inverting only once. Much larger economies are possible where the data are on a regular grid because the same configuration recurs many times. Not only does the variogram remain constant, but so also does matrix **A** for any given configuration. Each configuration requires only one matrix inversion. If sampling has been done on a square grid and the interpolation grid fits on to it with interval $1/r$ times that of the sampling grid then there are only r^2 possible configurations except near the edge of the map. Where variation is isotropic the spatial relations have a fourfold symmetry, so even fewer solutions are needed.

8.7   CASE STUDY

To illustrate the application of kriging to mapping we return to the analysis of exchangeable potassium (K) from Broom's Barn Farm. Since the distribution of K is skewed (skewness 2.04, Table 2.1) we transformed the values to common logarithms ($\log_{10} K$) which reduced the skewness to 0.39.

The variogram was computed on the transformed data, and the experimental semivariances were fitted best by a spherical function, equation (8.19), using a weighted least-squares approximation as described in Chapter 6. The resulting coefficients are $c_0 = 0.004\,66$, $c = 0.015\,15$

and a = 432 m. Figure 8.13 shows the experimental variogram (symbols) and the fitted spherical model (solid line). This function was then used for the kriging. We set the maximum radius of the neighbourhood to 400 m, and we set the minimum and maximum number of points to seven and 20 respectively. We kriged at intervals of 10 m, and for the block kriging our blocks were 50 m × 50 m. The estimated values and estimation variances have been mapped using Gsharp.

Figure 8.14 is a map of the punctual estimates. For it we deliberately placed the kriging grid over the sampling grid so that the sampling points lay on it to illustrate the nugget effect. The map is somewhat 'spotty' because we have kriged at the data points. The spatial pattern of $\log_{10} K$ is distinctly patchy, as we should expect from the spherical variogram; there are patches of large values and patches of small ones. The average extent of the patches is about 400 m.

In the alternative representation as a perspective diagram (Figure 8.15), the spots now appear as spectacular spikes, both above and below the surface. The reason is that at the sampling points punctual kriging returns the measured values there, whereas elsewhere it forms weighted averages of the data. The nugget variance in the variogram represents a discontinuity (Chapter 6), and this continues through to the kriging. Another way of viewing the effect is to consider the estimate as comprising two parts: the nugget variance and the continuous autocorrelated variation. Combining these two components produces the effect. The larger is the nugget variance as a proportion of the total variance the more pronounced this effect becomes, until when all of the variance is nugget the surface becomes flat between the sampling points.

Figure 8.16 is a map of the block estimates which has lost the 'spotty' appearance of Figure 8.14. Nevertheless, the same broad pattern in the distribution of $\log_{10} K$ is evident. The block kriged surface is smoother, and this is evident in the perspective diagram of this surface, Figure 8.17.

The estimation variances for punctual kriging are shown in Figure 8.18. In general they are much larger than those for the block kriging, see Figure 8.20, and the technique therefore appears much less precise. At the data points, however, they are zero. The perspective diagram (Figure 8.19), shows both features. Between the sampling points the nugget variance sets a minimum to the kriging variance, and at the sampling points the surface descends to zero.

The block kriging variances (Figures 8.20 and 8.21), are in general small, but they increase rapidly near the boundaries of the farm, beyond which there are no data, and similarly around the farm buildings (left above centre). The small ridge on the right of Figure 8.21 is an access road, again without any data along it, and the small hump on the lower left-hand side is where two data were lost.

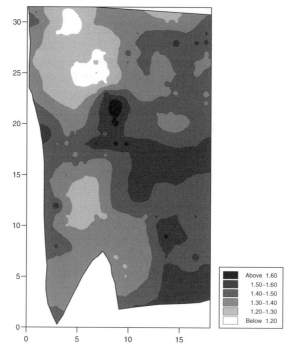

Figure 8.14 Map of exchangeable potassium, transformed to common logarithms, at Broom's Barn Farm, made by punctual kriging on a 10 m × 10 m grid that coincided with the sampling grid. The units are $\log_{10}(\text{mg K l}^{-1})$.

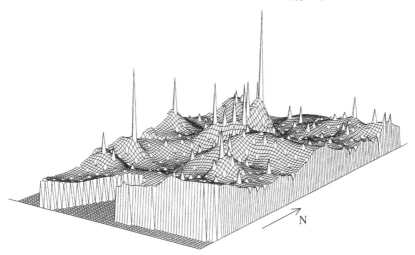

Figure 8.15 Perspective diagram of exchangeable potassium transformed to common logarithms at Broom's Barn Farm made by punctual kriging on a 10 m × 10 m grid that coincided with the sampling grid.

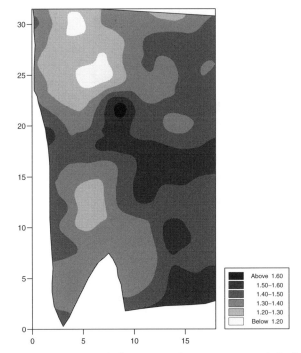

Figure 8.16 Map of $\log_{10}(\text{mg K l}^{-1})$ at Broom's Barn Farm made by kriging 50 m×
50 m blocks on a 10 m × 10 m grid.

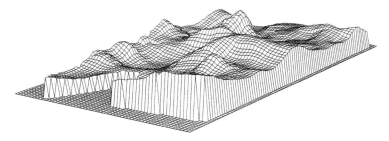

Figure 8.17 Perspective diagram of exchangeable potassium at Broom's Barn
Farm made by kriging 50 m × 50 m blocks on a 10 m × 10 m grid.

8.7.1 Summary

In practice exact interpolation might not be as attractive as one imag-
ines. Nevertheless, the effect of the nugget variance can be avoided either
by offsetting the kriging grid so that estimates are not made at a data
point or by omitting a data point when it coincides with the point to be
estimated.

We can use the maps or diagrams of the estimation variance as a guide
to the reliability of our estimates, but with caution. The reliability of krig-

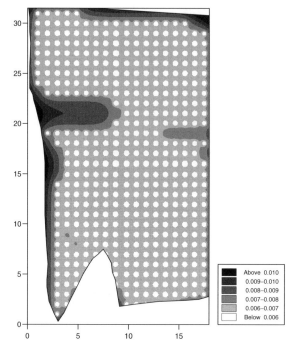

Figure 8.18   Map of the estimation variances of $\log_{10}(\text{mg K l}^{-1})$ at Broom's Barn Farm for punctual kriging.

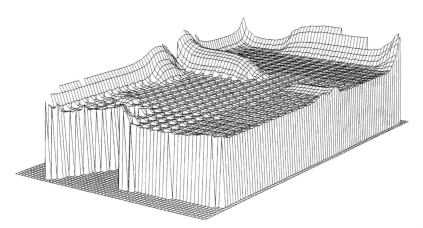

Figure 8.19   Perspective diagram of the estimation variances of $\log_{10}(\text{mg K l}^{-1})$ at Broom's Barn Farm for punctual kriging.

ing depends on how accurately the variation is represented by the chosen spatial model. If the nugget variance is overestimated then so will be the punctual estimation variances, and our estimates should be more reliable

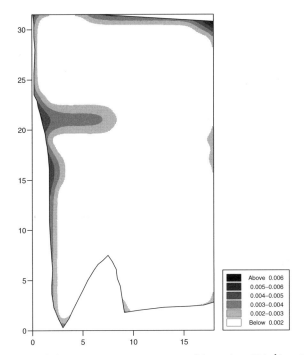

Figure 8.20 Map of the estimation variances of $\log_{10}(\text{mg K l}^{-1})$ at Broom's Barn Farm for block kriging over 50 m $\times$ 50 m blocks on a 10 m $\times$ 10 m grid.

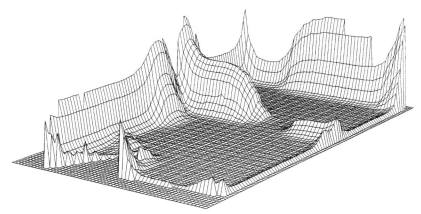

Figure 8.21 Perspective diagram of the estimation variances of exchangeable potassium at Broom's Barn Farm for block kriging over 50 m $\times$ 50 m blocks on a 10 m $\times$ 10 m grid.

than they appear to be. With block kriging the reverse can be the case, and we might imagine our estimates to be more reliable than they are. The block estimation variance comprises three terms, one of which is the within-block variance. The latter is estimated by integrating the variogram from $|\mathbf{h}| = 0$ to the limit of the block, see Figure 8.1(b). If the semivariance is overestimated at short lags then the within-block variance will also be overestimated, at least for small blocks the sides of which are less than the shortest sampling interval of the variogram. The estimates might therefore be less reliable than they appear. For larger blocks estimates should be reliable because the contribution to the within-block variance from the short lags will be a small proportion of the whole.

8.8   REGIONAL ESTIMATION

In the limit we can think of the whole region, R, of interest as a single large block for which we could estimate the mean of Z, $\hat{Z}(R)$, by including all the data. In classical estimation this is precisely what we do, giving all data the same weight, see equation (2.34). The solution takes no account of known spatial correlation, and kriging should do better by assigning differential weights.

We assume first that $Z(\mathbf{x})$ is second-order stationary with mean μ and variance σ^2. As R increases so the average distance between pairs of points in it increases, and the average semivariances, $\overline{y}(\mathbf{x}_i, B)$, in equation (8.11) approach σ^2, the sill of the variogram. If the distance across R is much larger than the effective range of the variogram then the $\overline{y}(\mathbf{x}_i, B)$ will be so close to σ^2 that the two can be taken as equal. The kriging system (8.11) can therefore be rewritten

$$\sum_{i=1}^{N} \lambda_i y(\mathbf{x}_i, \mathbf{x}_j) + \psi(R) = \hat{\sigma}^2 \qquad \text{for all } j,$$

$$\sum_{i=1}^{N} \lambda_i = 1. \tag{8.22}$$

The kriging weights are found in the usual way, and the kriging variance, from equation (8.16), is

$$\sigma^2(R) = \mathbf{b}^{\mathsf{T}}\boldsymbol{\lambda} - \sigma^2$$

$$= \sigma^2 \sum_{i=1}^{N} \lambda_i + \psi(R) - \overline{y}(R, R). \tag{8.23}$$

The sum $\sum_{i=1}^{N} \lambda_i = 1$, and so we have that

$$\sigma^2(R) = \psi(R). \tag{8.24}$$

Since $Z(\mathbf{x})$ must be second-order stationary, the covariances exist, and the kriging system is usually expressed in terms of covariances:

$$\sum_{i=1}^{N} \lambda_i C(\mathbf{x}_i, \mathbf{x}_j) - \psi(R) = 0 \qquad \text{for all } j,$$

$$\sum_{i=1}^{N} \lambda_i = 1,$$

(8.25)

from which it follows immediately that $\sigma^2(R) = \mathbf{b}^T\boldsymbol{\lambda} + \psi(R)$.

Kriging the mean is undoubtedly attractive from a theoretical point of view. Unfortunately there are reasons why the approach cannot or should not be pursued.

1. It is unwise to assume that a property, which is locally stationary in the mean and semivariances, maintains that stationarity throughout a large region.

2. The experimental variogram is usually known accurately only for the first few lags; it almost certainly will not be well estimated for lags approaching the distance across a large region.

3. A large sample could produce kriging matrices that are too large to invert or that become unstable.

A practical alternative that avoids the difficulties is to divide the region into small rectangular blocks or strata, estimate the mean in each by kriging, and then compute the average of the estimates. If for some reason the blocks are not all of the same size then their estimates can be weighted according to their areas. For a region, R, divided into n blocks, B_i, $i = 1, 2, \ldots, n$, of area H_i, the global mean, $Z(R)$, is estimated by

$$\hat{Z}(R) = \sum_{i=1}^{n} H_i \hat{Z}(B_i) \Big/ \sum_{i=1}^{n} H_i,$$

(8.26)

where $\hat{Z}(B_i)$ is the kriged estimate of Z within the ith block. If the blocks are of equal size then the H_i cancel, and $\hat{Z}(R) = \sum_{i=1}^{n} \hat{Z}(B)/n$.

A problem arises in calculating the estimation variance. The error in the global average equals the sum of the errors in the local estimates:

$$\hat{Z}(R) - Z(R) = \sum_{i=1}^{n} H_i \{\hat{Z}(B_i) - Z(B_i)\} \Big/ \sum_{i=1}^{n} H_i.$$

(8.27)

The estimation variance, $\sigma^2(R) = E[\{\hat{Z}(R) - Z(R)\}^2]$, cannot be estimated without bias by a simple sum, however, because the estimates in the neighbouring blocks are not independent; some of the data from which they are computed are common. We can solve the problem by considering the error that results from using the value at a sampling point to estimate the average value over the portion of the region that is nearer to it than to any other, i.e. for its Thiessen polygon or Dirichlet tile. For a rectangular grid each polygon is a rectangle with an observation at its centre, $\mathbf{x}_c$, and sides equal to the sampling intervals along the principal axes of the grid. The variance of the estimate of its average is

$$\sigma^2(B) = 2\overline{\gamma}(\mathbf{x}_c, B) - \overline{\gamma}(B, B), \tag{8.28}$$

where $\overline{\gamma}(\mathbf{x}_c, B)$ is the average semivariance between the centre and all other points in the rectangle, and $\overline{\gamma}(B, B)$ is the variance within the polygon. Since the estimated values for these rectangles are $\hat{Z}(B_i)$, $i = 1, 2, \ldots$, n, the average for the region is approximately

$$\hat{Z}_B(R) = \frac{1}{n} \sum_{i=1}^{n} \hat{Z}(B_i). \tag{8.29}$$

The error of this estimate is approximately $Z(R) - \hat{Z}_B(R)$, and the corresponding variance of the regional mean is

$$E[\{Z(R) - \hat{Z}_B(R)\}^2] \approx \frac{1}{n^2} \sum_{i=1}^{n} E[\{Z(B_i) - Z(\mathbf{x}_i)\}^2]$$

$$= \frac{1}{n} \sigma^2(B). \tag{8.30}$$

The approximation improves as n increases.

Thus the error of the regional estimate depends on the variances within small rectangular blocks, and these are likely to be much smaller than the variance within the entire region.

8.9   SIMPLE KRIGING

Sometimes we know the mean of a random variable from previous experience or we can assume it from the nature of the problem. In these circumstances we should use that knowledge to improve our estimates, and we can do so by 'simple kriging'. Our kriged estimate is still a linear sum, but now incorporating the mean, μ, of the process, which must be second-order stationary. Prediction by simple kriging is not an option for processes that are intrinsic only, a variogram with an upper bound is needed. For punctual kriging the equation is

$$\hat{Z}_{SK}(\mathbf{x}_0) = \sum_{i=1}^{N} \lambda_i z(\mathbf{x}_i) + \left\{ 1 - \sum_{i=1}^{N} \lambda_i \right\} \mu. \tag{8.31}$$

The λ_i are the weights, as before, but they are no longer constrained to sum to 1. The unbiasedness is assured by including the second term on the right-hand side of equation (8.31). The weights are found by solving

$$\sum_{N=1}^{N} \lambda_i \gamma(\mathbf{x}_i, \mathbf{x}_j) = \gamma(\mathbf{x}_0, \mathbf{x}_j) \quad \text{for } j = 1, 2, \ldots, N. \tag{8.32}$$

There is no Lagrange multiplier: there are only N equations in N unknowns. The kriging variance is given by

$$\sigma_{SK}^2(\mathbf{x}_0) = \sum_{i=1}^{N} \lambda_i \gamma(\mathbf{x}_i, \mathbf{x}_0). \tag{8.33}$$

As with ordinary kriging the technique can be generalized for blocks, B, larger than the supports of the sample by replacing the $\gamma(\mathbf{x}_0, \mathbf{x}_j)$ on the right-hand side of equation (8.32) by the averages $\bar{\gamma}(B, \mathbf{x}_j)$. Also, N, the total size of the sample, can usually be replaced by $n \ll N$ data in close proximity to $\mathbf{x}_0$ or B.

In general, the variances obtained by simple kriging are somewhat smaller than those from ordinary kriging, and we might think that we could improve the predictions by introducing the mean estimated from the data, $\hat{\mu}$. Wackernagel (1995) shows that if the kriged mean is used, i.e. by putting $\hat{\mu} = \hat{Z}(R)$, it results in the ordinary kriging predictor with variance

$$\sigma_{OK}^2(\mathbf{x}_0) = \sigma_{SK}^2(\mathbf{x}_0) + \left\{ 1 - \sum_{i=1}^{N} \lambda_i^{SK} \right\}^2 \psi(R). \tag{8.34}$$

In words, the ordinary kriging variance is the sum of the simple kriging variance plus the variance arising from the estimate of the mean. There is nothing to be gained by taking this approach because there is no more information. If the mean is estimated from many data, as will usually be the case, then $\psi(R)$ will be small in relation to $\sigma_{SK}^2(\mathbf{x}_0)$, and provided the simple kriging weights are close to 1 then the second term on the right-hand side of equation (8.34) is likely to be very small indeed.

8.10 LOGNORMAL KRIGING

A more common situation in the environmental sciences, and in mining and petroleum engineering too, is that the data are often markedly skewed, and non-normal. As mentioned in Chapter 5, the variogram is sensitive to strong positive skewness because a few exceptionally large values contribute to so many squared differences. Such skewness can often be removed and the variances stabilized by taking logarithms. If by transforming to logarithms the distribution is made near-normal then it is said to be lognormal. This leads to lognormal kriging.

The data $z(\mathbf{x}_1), z(\mathbf{x}_2), \ldots$, are transformed to their corresponding natural logarithms, say $y(\mathbf{x}_1), y(\mathbf{x}_2), \ldots$, which represent a sample from the random variable $Y(\mathbf{x}) = \ln Z(\mathbf{x})$, which is assumed to be second-order stationary. The variogram of $Y(\mathbf{x})$ is computed and modelled and then used with the transformed data to estimate Y at the target points or blocks by either ordinary or simple kriging. The estimated values are in logarithms.

For some purposes, as for example at Broom's Barn Farm where an index of soil fertility is wanted, the logarithms can serve well. However, in many other disciplines, such as mining, exploration geochemistry, and pollution monitoring, surveyors want estimates expressed in the original units, and the logarithms must be transformed back to concentration.

The back-transformation is fairly straightforward. If we denote the kriged estimate of the natural logarithm at $\mathbf{x}_0$ as $\hat{Y}(\mathbf{x}_0)$ and its variance as $\sigma^2(\mathbf{x}_0)$, then the formulae for the back-transformation of the estimates are, for simple kriging,

$$\hat{Z}_{SK}(\mathbf{x}_0) = \exp\{\hat{Y}_{SK}(\mathbf{x}_0) + \sigma_{SK}^2(\mathbf{x}_0)/2\}, \tag{8.35}$$

and for ordinary kriging,

$$\hat{Z}_{OK}(\mathbf{x}_0) = \exp\{\hat{Y}_{OK}(\mathbf{x}_0) + \sigma_{OK}^2(\mathbf{x}_0)/2 - \psi\}, \tag{8.36}$$

where ψ is the Lagrange multiplier in the ordinary kriging. The estimation variance of $Z(\mathbf{x}_0)$ for simple kriging is

$$\mathrm{var}_{SK}[\hat{Z}(\mathbf{x}_0)] = \mu^2 \exp(\sigma_{SK}^2)[1 - \exp\{-\sigma_{SK}^2(\mathbf{x}_0)/2\}], \tag{8.37}$$

where μ is the mean of $Z(\mathbf{x})$. We cannot obtain an unbiased back-transform of the ordinary kriging variance because the mean, μ, is not known.

In many fields of application people prefer to work with common logarithms. The variogram of $\log_{10} Z(\mathbf{x})$ replaces that of $\ln Z(\mathbf{x})$, and the back-transform for ordinary kriging, is given by

$$\hat{Z}(\mathbf{x}_0) = \exp\{\hat{Y}(\mathbf{x}_0) \times \ln 10 + 0.5\sigma_Y^2(\mathbf{x}_0) \times (\ln 10)^2\}. \tag{8.38}$$

Journel and Huijbregts (1978) point out that expression (8.35) for the back-transformation is sensitive to departures from lognormality and that in consequence the estimates of Z can be biased. They suggest a check for bias by comparing the mean of the estimates, $\hat{Z}$, with the mean of the data, $z(\mathbf{x}_i), i = 1, 2, \ldots, N$. If we denote the ratio of the means, $\mathrm{mean}[\hat{Z}] : \bar{z}$, by Q then we modify equation (8.35) to

$$\hat{Z}_{SK}(\mathbf{x}_0) = Q \exp\{\hat{Y}_{SK}(\mathbf{x}_0) + \sigma_{SK}^2(\mathbf{x}_0)/2\}, \tag{8.39}$$

or equation (8.36) in like manner if we have used ordinary kriging. In our experience Q has always been so close to 1 that we have not needed the elaboration. Figure 8.22 shows the back-transformed values of the block-kriged estimates of $\log_{10} K$.

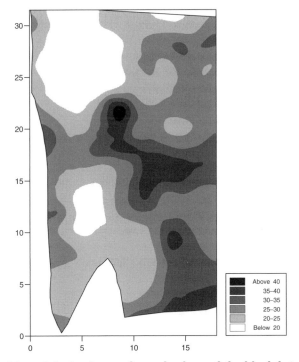

Figure 8.22   Map of the back-transformed values of the block kriged estimates of $\log_{10} K$ at Broom's Barn Farm.

8.11   UNIVERSAL KRIGING

We mentioned in Chapter 5 that some spatial processes comprise both stochastic and deterministic components, and we represented them by the model

$$Z(\mathbf{x}) = \sum_{k=0}^{K} a_k f_k(\mathbf{x}) + \varepsilon(\mathbf{x}).\qquad(8.40)$$

In this equation the (non-stationary) drift is represented in the first term on the right-hand side by a set of functions, $f_k(\mathbf{x})$, $k = 0, 1, \ldots, K$, of our choosing and unknown coefficients a_k. The functions are usually simple polynomials of order 1 or 2, and $f_0(\mathbf{x}) = 1$. The second term, $\varepsilon(\mathbf{x})$, is the stochastic component, the variogram of which we shall wish to use for kriging. We described in Chapter 5 how to compute a variogram of the residuals from the drift from regularly spaced data. Now we give the formulae for kriging using that variogram. Matheron (1969) called the technique 'universal kriging', though, as Webster and Burgess (1980) pointed out, it is far from universal in its applicability. It is now more often and more aptly called 'kriging in the presence of drift'.

As above, we seek a linear sum of the data. For punctual estimates this is given by

$$\hat{Z}(\mathbf{x}_0) = \sum_{k=0}^{K} \sum_{i=1}^{N} a_k \lambda_i f_k(\mathbf{x}_i). \tag{8.41}$$

Our estimator $\hat{Z}(\mathbf{x})$ is unbiased if and only if

$$\sum_{i=1}^{N} \lambda_i f_k(\mathbf{x}_i) = f_k(\mathbf{x}_0) \qquad \text{for all } k = 1, 2, \ldots, K. \tag{8.42}$$

We must now find the weights that minimize the variance, subject to this condition, and we do so by solving the following system of equations:

$$\sum_{i=1}^{N} \lambda_i \gamma(\mathbf{x}_i, \mathbf{x}_j) + \psi_0 + \sum_{k=1}^{K} \psi_k f_k(\mathbf{x}_j) = \gamma(\mathbf{x}_0, \mathbf{x}_j) \qquad \text{for all } j = 1, 2, \ldots, N,$$

$$\sum_{i=1}^{N} \lambda_i = 1, \tag{8.43}$$

$$\sum_{i=1}^{N} \lambda_i f_k(\mathbf{x}_i) = f_k(\mathbf{x}_0) \qquad \text{for all } k = 1, 2, \ldots, K.$$

The values $\gamma(\mathbf{x}_i, \mathbf{x}_j)$ are the semivariances of the residuals between the data points $\mathbf{x}_i$ and $\mathbf{x}_j$, and the $\gamma(\mathbf{x}_0, \mathbf{x}_j)$ are the semivariances between the target and the data points. The functions $f_k(\mathbf{x})$, $k = 0, 1, \ldots, K$, refer to an origin, $\mathbf{x} = \mathbf{0}$, i.e. $\{x_1 = 0, x_2 = 0\}$, for the target point $\mathbf{x}_0$. For a linear drift there are three, i.e. $K = 3$, with values

$$f_0 = 1, \qquad f_1 = x_1, \qquad f_2 = x_2.$$

For quadratic drift $K = 6$, and the three additional functions have values

$$f_4 = x_1^2, \qquad f_5 = x_1 x_2, \qquad f_6 = x_2^2.$$

In addition, there are now three Lagrange multipliers, ψ_0, ψ_1 and ψ_2, for the linear drift and three more, ψ_3, ψ_4 and ψ_5, for quadratic drift.

The universal kriging system, like that for ordinary kriging, is a set of linear equations which we can represent in matrix notation by

$$\mathbf{A}\boldsymbol{\lambda} = \mathbf{b}, \tag{8.44}$$

as in equation (8.13). Now, however, the matrix $\mathbf{A}$ and the vectors $\boldsymbol{\lambda}$ and $\mathbf{b}$ are augmented with functions of the spatial positions of the data points and of the target. They are as follows:

$$
\mathbf{A} =
\begin{bmatrix}
\gamma(\mathbf{x}_1,\mathbf{x}_1) & \gamma(\mathbf{x}_1,\mathbf{x}_2) & \cdots & \gamma(\mathbf{x}_1,\mathbf{x}_N) & 1 & f_1(\mathbf{x}_1) & f_2(\mathbf{x}_1) & \cdots & f_K(\mathbf{x}_1) \\
\gamma(\mathbf{x}_2,\mathbf{x}_1) & \gamma(\mathbf{x}_2,\mathbf{x}_2) & \cdots & \gamma(\mathbf{x}_2,\mathbf{x}_N) & 1 & f_1(\mathbf{x}_2) & f_2(\mathbf{x}_2) & \cdots & f_K(\mathbf{x}_2) \\
\vdots & \vdots & \cdots & \vdots & \vdots & \vdots & \vdots & \cdots & \vdots \\
\gamma(\mathbf{x}_N,\mathbf{x}_1) & \gamma(\mathbf{x}_N,\mathbf{x}_2) & \cdots & \gamma(\mathbf{x}_N,\mathbf{x}_N) & 1 & f_1(\mathbf{x}_N) & f_2(\mathbf{x}_N) & \cdots & f_K(\mathbf{x}_N) \\
1 & 1 & \cdots & 1 & 0 & 0 & 0 & \cdots & 0 \\
f_1(\mathbf{x}_1) & f_1(\mathbf{x}_2) & \cdots & f_1(\mathbf{x}_N) & 0 & 0 & 0 & \cdots & 0 \\
f_2(\mathbf{x}_1) & f_2(\mathbf{x}_2) & \cdots & f_2(\mathbf{x}_N) & 0 & 0 & 0 & \cdots & 0 \\
\vdots & \vdots & \cdots & \vdots & \vdots & \vdots & \vdots & \cdots & \vdots \\
f_K(\mathbf{x}_1) & f_K(\mathbf{x}_2) & \cdots & f_K(\mathbf{x}_N) & 0 & 0 & 0 & \cdots & 0
\end{bmatrix},
$$

$$
\boldsymbol{\lambda} =
\begin{bmatrix}
\lambda_1 \\ \lambda_2 \\ \vdots \\ \lambda_N \\ \psi_0 \\ \psi_1 \\ \psi_2 \\ \vdots \\ \psi_K
\end{bmatrix}, \qquad
\mathbf{b} =
\begin{bmatrix}
\gamma(\mathbf{x}_1,\mathbf{x}_0) \\ \gamma(\mathbf{x}_2,\mathbf{x}_0) \\ \vdots \\ \gamma(\mathbf{x}_N,\mathbf{x}_0) \\ 1 \\ f_1(\mathbf{x}_0) \\ f_2(\mathbf{x}_0) \\ \vdots \\ f_K(\mathbf{x}_0)
\end{bmatrix}.
$$

As in ordinary kriging, $\mathbf{A}$ is inverted, and the weights and the Lagrange multipliers are obtained as

$$
\boldsymbol{\lambda} = \mathbf{A}^{-1}\mathbf{b}. \tag{8.45}
$$

The weights are inserted into equation (8.41), and the kriging variance is given by

$$
\sigma^2_{\mathrm{UK}} = \mathbf{b}^{\mathrm{T}}\boldsymbol{\lambda}. \tag{8.46}
$$

Thus universal kriging looks remarkably like ordinary kriging, and like ordinary kriging the procedure is automatic once you have a satisfactory function for the variogram. The difficult task is obtaining such a function in the presence of drift.

8.12 OPTIMAL SAMPLING FOR MAPPING

From equations (8.2) and (8.4) it is evident that the kriging weights depend on the configuration of the sampling points in relation to the target point or block and on the variogram. They do not depend at all on the observed values at those points. The same applies to the kriging variances, see equation (8.2). Therefore, if we know the variogram then we can determine the kriging errors for any sampling configuration *before* doing the sampling, and we can design a sampling scheme to meet a specified tolerance or precision.

In general, mapping is most efficient if survey is done on a regular grid in the sense that the maximum kriging error is minimized. Where there is spatial dependence the information from an observation pertains to an area surrounding it, and specifically to the neighbourhood within its range if the variable is second-order stationary. If the neighbourhoods of two observations overlap then information is duplicated to some extent. Any kind of clustering of points, such as arises with random sampling, means that information can be replicated while elsewhere there is under-representation or even big gaps. Redundancy can be minimized by placing the sampling points as far away from their neighbours as possible for a given sampling density. This approach also minimizes the area that is under-represented. Triangular configurations are the most efficient in this respect. For a grid with one node per unit area neighbouring sampling points are 1.0746 units of distance apart, and no point is more than 0.6204 units away from another. We denote this maximum distance d_{max}. Rectangular grids have some neighbours that are closer and others that are further away. For a square grid with one node per unit area the sampling interval is 1, and $d_{max} = 1/\sqrt{2} = 0.7071$. For a hexagonal grid with unit sampling density $d_{max} = 0.8772$. From this we should expect triangular sampling configurations to be the most efficient. Matérn (1960) and Dalenius *et al.* (1961) showed that where the variogram is exponential the triangular grid is optimal for estimating the mean of a region, and in most circumstances with bounded variograms that have finite ranges. The same is also true if the variogram is unbounded. In certain restricted circumstances with variograms with a finite range, a hexagonal grid can be the most efficient (Yfantis *et al.*, 1987). In general, however, rectangular grids are preferred because they are easier to work with in the field. Figure 8.23(a) shows that the difference in precision between a triangular configuration and a square one is small, and that we can choose the type of grid that we prefer to work with.

The variogram then enables us to optimize the sampling interval to estimate both the regional mean and local values for mapping. For estimation by kriging, or indeed any other method of interpolation, the distances between neighbouring sampling points should be well within the correlation range. As we have seen above, if they are beyond the range then kriging simply returns the mean of the points in the neighbourhood.

The kriging errors are not the same everywhere. With punctual kriging there is no error at the sampling points, see Figure 8.11(c), and, in general, the further a target point is from the data the larger the error. If we sample

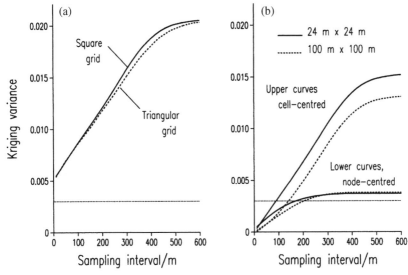

Figure 8.23 Graphs of kriging variance against sampling interval to map exchangeable potassium at Broom's Barn Farm. (a) Punctual kriging variances for square and triangular grids. (b) Kriging variances for blocks of 24 m × 24 m and 100 m × 100 m; the upper two curves are for blocks centred at centres of grid cells, and the lower two, which are distinguishable only at fairly short sampling intervals, are for blocks centred over grid nodes. The horizontal lines are for a tolerable kriging variance of 0.003. This is approximately equivalent to a 90% confidence interval of 10 mg l^{-1} at the critical concentration of 25 mg kg^{-1} soil.

on a regular grid we minimize d_{max}, which is the distance between a target point at the centre of a grid cell and its nearest sampling point on the grid. We also minimize the maximum kriging error, except near the margins of the map.

8.12.1   Isotropic variation

Following Burgess *et al.* (1981) and McBratney *et al.* (1981), the kriging equations can be solved to design an optimal sampling scheme. For punctual kriging we solve equations (8.9), and determine the kriging variances and errors by equation (8.10) at the centres of grid cells for a range of sampling intervals. The variances can then be plotted against the grid spacing. If we have in mind a maximum variance or error that we can tolerate then we can draw a horizontal line across the graph until it meets the maximum kriging variance. A perpendicular from this point gives the optimal sample spacing.

To illustrate the procedure we use the variogram $\log_{10} K$ for Broom's Barn Farm (Figure 8.13 and Table 6.1). Figure 8.23(a) shows the maximum punctual kriging variance for square and triangular grids. Note that the difference between the curves for the square and triangular grids is not nearly as large as the 12% difference in d_{max} for the two grids. The line drawn across the graph at 0.003 is the kriging variance on the logarithmic scale that is approximately equivalent to a 90% confidence interval of 10 mg l^{-1} at the deficiency threshold of 25 mg l^{-1}. The kriging variances are large, and many exceed this tolerance.

Figure 8.23(a) illustrates two other features of punctual kriging. First, if we set the maximum tolerable kriging variance at 0.003 then it is impossible to design a satisfactory sampling scheme because we cannot diminish σ_{max}^2 to less than the nugget variance, 0.004 66. Second, σ_{max}^2 increases to a maximum at which it flattens. This maximum is somewhat larger than the sill of the variogram; in fact it is the sill plus the Lagrange multiplier, ψ of equation (8.9). Once d_{max} exceeds the range of the variogram, 432 m in this case, all the semivariances in the kriging system are equal, as are the weights, as we saw in the example above. The additional quantity ψ represents the additional uncertainty of predicting the value at a place from only local data.

The same reasoning and procedure apply to block kriging (equation (8.11)). However, it is less straightforward, and the result depends on the block size. For blocks of side much smaller than the sampling interval the kriging variance will be largest when the blocks are in the centres of grid cells. As the block is increased in size the kriging variance decreases (contrast the 24 m × 24 m blocks with the 100 m × 100 m blocks in Figure 8.23(b)). Consider, however, a block centred on a grid node. If the block is no larger than the sample support this is effectively punctual kriging and the estimation variance is zero. As the block size increases, its kriging variance initially increases because the dominant effect of the observation at its centre declines. Only when it is big enough for the nearest neighbours to be more influential does the kriging variance start to decline. This difference in the configuration has another important effect. As the block increases in size the weights of the sampling points nearest its centre decrease, whereas the weights of those further away increase (see Figure 8.9). A block size is eventually reached at which its estimation variance equals that for a block centred in a grid cell. If the block size becomes larger still the estimation variance can be greater than that of a block of the same size centred in a grid cell. Therefore, for block kriging one must decide where to determine the kriging variances, i.e. whether for blocks centred on grid cells or ones centred on grid nodes. The position at which the kriging variance is greatest for a given block size is the one to choose. Burgess *et al.* (1981) describe these effects in detail.

In Figure 8.23(b) the kriging variances for blocks centred at the cell centres and grid nodes are plotted against distance for a square grid for blocks of side 24 m and 100 m. At the chosen tolerance the horizontal line intersects the graph of the variances for blocks about 80 m apart for blocks of side 24 m and about 130 m apart for 100 m blocks.

For block kriging of potassium at Broom's Barn Farm, the results suggest that sampling might have been denser than necessary for mapping.

Using the variogram and the kriging equations, it is possible to design a new survey to be optimal in the sense that sampling is just sufficiently intense to meet the specified tolerance. Near the margins of the region some modifications might be needed if sampling cannot be extended outside it because the variance increases at the margin (see Figures 8.19 and 8.21); sampling would need to be increased near the margin to keep within the tolerance.

We can also use this approach if we feel that part of a region is undersampled. We can see whether adding further points will increase the precision before sampling more. Also, if we have a network of stations for monitoring rainfall or pollutants in ground water the effect of adding stations, moving them or removing them can be assessed. This is what McCullagh (1976) did with the Trent telemetry network. Barnes (1989) used different strategies to optimize the placement of a new sampling station—depending on whether the need was to improve the worst situation or to diminish the estimation variance on average.

This approach allows sampling to be optimized in the sense of minimizing effort.

8.12.2 Anisotropic variation

One can take anisotropy into account when planning sampling. The grid spacing is adjusted so that the sampling is more intense in the direction of minimum continuity, i.e. the direction with the maximum rate of spatial change, than in other directions. The problem is to keep within the specified tolerable error for least effort. The optimum solution depends on the form of the anisotropy. The one that we illustrate is for strict geometric anisotropy (Burgess *et al.*, 1981).

Consider the linear variogram

$$\gamma(h, \theta) = \Omega(\theta)|\mathbf{h}|, \tag{8.47}$$

in which $\Omega(\theta)$ is the sinusoidal function

$$\Omega(\theta) = \sqrt{A^2\cos^2(\theta - \phi) + B^2\sin^2(\theta - \phi)}. \tag{8.48}$$

In this equation ϕ is the direction of maximum variation, A is the gradient of the variogram in that direction, and B is the gradient in the perpendicular direction, $\phi + \pi/2$. When $\theta = \phi$, equation (8.47) reduces to

$$\gamma_1(h) = Ah, \tag{8.49}$$

and when $\theta = \phi + \pi/2$ it becomes

$$y_2(h) = Bh. \tag{8.50}$$

As above, we can define an anisotropy ratio R:

$$R = A/B = y_1(h)/y_2(h). \tag{8.51}$$

The semivariance in direction ϕ at any lag h is thus equal to the semivariance at lag Rh in the direction $\phi + \pi/2$:

$$y_1(h) = y_2(Rh). \tag{8.52}$$

Using equation (8.52), the most economical sampling scheme is found as follows. The problem is treated as though variation were isotropic with the variogram $y_1(h)$. The sampling interval d is found in exactly the same way as for the square grid. This then becomes the sampling interval in direction ϕ. The anisotropy is taken into account by making the sampling interval in the perpendicular direction, $\phi + \pi/2$, equal to Rd.

8.13   OTHER KINDS OF KRIGING

Kriging is a generic term that covers a range of least-squares methods of spatial prediction. Ordinary kriging of a single variable, as described above, is the most robust and the one most used.

Simple kriging is rather little used as it stands because we usually do not know the mean. It finds application in other forms, such as indicator and disjunctive kriging, in which the data are transformed to have known means.

Universal kriging might be used more than it is if it were easier to estimate the variogram of the residuals; that difficulty is undoubtedly a stumbling block.

Indicator kriging (see also Chapter 10) is a non-linear, non-parametric form of kriging in which continuous variables are converted to binary ones (indicators). It is becoming popular because it can handle distributions of almost any kind, and empirical cumulative distributions of estimates can be computed and thereby provide confidence limits on them. It can also accommodate 'soft' qualitative information to improve prediction. Goovaerts (1997) describes the technique at length.

Disjunctive kriging (see also Chapter 10) is also a non-linear method of kriging, but it is strictly parametric. It is valuable for decision-making because the probabilities of exceeding or not exceeding a predefined threshold are determined in addition to the kriged estimates. Rivoirard (1994) describes the method most lucidly.

Ordinary cokriging is the extension of ordinary kriging of a single variable to two or more variables. There must be some coregionalization among the variables for it to be profitable. It is particularly useful if some property that can be measured cheaply at many sites is spatially correlated with one or more others that are expensive to measure and are measured at many fewer sites. The more sparsely sampled property can be estimated with more precision by cokriging using the spatial information from the more intensely measured one. Myers (1982) describes the theory and McBratney and Webster (1983), Vauclin *et al.* (1983) and Leenaers *et al.* (1990) provide examples. We describe the method in Chapter 9 and show its potential for diminishing the kriging variance.

Probability kriging was proposed by Sullivan (1984) because indicator kriging does not take into account the proximity of a value to the threshold, but only its position. It uses the rank order for each value, $z(\mathbf{x})$, normalized to 1, as the secondary variable to estimate the indicator by cokriging. Chilès and Delfiner (1999) and Goovaerts (1997) describe the method briefly.

Bayesian kriging was introduced by Omre (1987) for situations in which there is some prior knowledge about the drift. It is intermediate between simple kriging, used when there is no drift, and universal kriging where there is known to be drift. The kriging equations are those of simple kriging, but with non-stationary covariances (Chilès and Delfiner, 1999).

8.14 CROSS-VALIDATION

In Chapter 6 we fitted models by minimizing the deviations between the observed semivariances and the ones expected from the model, and we chose finally from among different kinds of model those for which the squared deviations were least on average. We weighted the experimental values in proportion to the numbers of pairs contributing to them, but we paid no regard to the lag except incidentally when we refined the weighting as a function of the expected value. This is not necessarily the best for kriging because points near the target point or block get more weight than more distant ones. So we should really like the variogram to be accurate at short lags, if necessary at the expense of less accuracy at longer lags. But how should we choose?

One way of choosing between competing models is to use them for kriging and see how well they perform. This may be done rigorously by having a separate set of sample data against which to compare kriged estimates. Except in research studies this is a waste of information, and validation usually is done by a process known as 'cross-validation'. It works as follows:

1. An experimental variogram is computed from the whole set of sample data, and plausible models are fitted to it.

Table 8.4 Mean error (ME), mean squared error (MSE), and mean squared deviation ratio (MSDR) for ordinary kriging $\log_{10} K$ using five models. The mean squared residuals are added for comparison.

Model	ME	MSE	MSDR	Mean squared residual
Circular	0.000 321	0.007 739	1.010	0.000 172
Spherical	0.000 327	0.007 639	1.044	0.000 155
Pentaspherical	0.000 346	0.007 584	1.081	0.000 248
Exponential	0.000 682	0.007 314	1.232	0.001 054
Power function	0.000 726	0.007 465	0.184	0.003 295

2. For each model, Z is estimated from the data and the model by kriging at each sampling point in turn after excluding the sample value there. The kriging variance is also calculated.

3. Three diagnostic statistics are calculated from the results:

   (a) the mean deviation or mean error, ME, given by

$$\text{ME} = \frac{1}{N} \sum_{i=1}^{N} \{z(\mathbf{x}_i) - \hat{z}(\mathbf{x}_i)\}; \tag{8.53}$$

   (b) the mean squared deviation or mean squared error, MSE:

$$\text{MSE} = \frac{1}{N} \sum_{i=1}^{N} \{z(\mathbf{x}_i) - \hat{z}(\mathbf{x}_i)\}^2; \tag{8.54}$$

   and

   (c) the mean squared deviation ratio, MSDR, computed from the squared errors and kriging variances, $\hat{\sigma}^2(\mathbf{x})$, by

$$\text{MSDR} = \frac{1}{N} \sum_{i=1}^{N} \frac{\{z(\mathbf{x}_i) - \hat{z}(\mathbf{x}_i)\}^2}{\hat{\sigma}^2(\mathbf{x}_i)}. \tag{8.55}$$

The mean error should ideally be 0 because kriging is unbiased. The calculated ME, however, is a weak diagnostic because kriging is insensitive to inaccuracies in the variogram. We want the MSE to be small, of course. If the model for the variogram is accurate then the MSE should equal the kriging variance; and so the MSDR should be 1.

Let us see how the models for $\log_{10} K$ at Broom's Barn Farm compare in this test. The three test criteria are listed in Table 8.4 for the five models summarized in Table 6.1, from which we have transferred the mean square residuals for comparison.

The first three models in the table, the circular, spherical and pentaspherical, have similar values for each of the three diagnostics. The MSDRs suggest that the kriging variances progressively underestimate the true estimation variances in that sequence, though not seriously. The MSE for the exponential model looks a little worrying, and we see that its mean squared residual is substantially larger than that of the first three models. The power function clearly performs poorly on the cross-validation with an MSDR of only 0.18. The kriging variance grossly exaggerates the true estimation variance. The mean squared residual tells the same story; that of the power function is quite the largest. Figure 6.12 suggests that its MSDR is so small because the model values exceed the observed ones at the short lags between the data and the target points, which are the ones that dominate the kriging systems.

8.14.1 Scatter and regression

Another way of examining the behaviour of kriging is to plot the scatter of the true values against their estimates. We should like the two to be the same, but perfection of this kind is elusive in nature. The best we can expect is that our estimator is conditionally unbiased, by which we mean

$$E[Z(\mathbf{x}_0) \mid \hat{Z}(\mathbf{x}_0)] = \hat{Z}(\mathbf{x}_0). \qquad (8.56)$$

From this it follows that the regression of $Z(\mathbf{x}_0)$ on $\hat{Z}(\mathbf{x}_0)$ is 1, therefore the covariance between the true values and their estimates must equal the variance of the estimates.

Armstrong (1998) shows that the above holds for simple kriging. For ordinary kriging, however, the variance of the estimates includes the Lagrange multiplier, and so the regression coefficient is somewhat less than 1.

Figure 8.24 illustrates the situation in which the true values are plotted against their estimates for $\log_{10} K$ at Broom's Barn Farm. The scatter forms an elliptical cloud with a few points lying outside it. The ellipse itself is a probability 'contour' (see Chapter 2) drawn to include all but a few of the points. Its diameters are proportional to the standard deviations along the principal axes, the longer of which is drawn with a dashed line. They and the orientation have been estimated by a principal component analysis. The regression of $Z(\mathbf{x}_0)$ on $\hat{Z}(\mathbf{x}_0)$ is the 1 : 1 line, the diagonal joining the corners of the frame and passing through the points where the vertical tangents touch the ellipse. The actual regression coefficients for simple and ordinary kriging estimated in this way are 1.035 and 1.024, respectively. They are barely distinguishable from 1. Like the mean error, this regression is a poor diagnostic because the kriged estimates are so insensitive to the model.

Figure 8.24 shows another feature of kriging. The long axis of the ellipse is oriented at about 56° from the horizontal; it is substantially more than

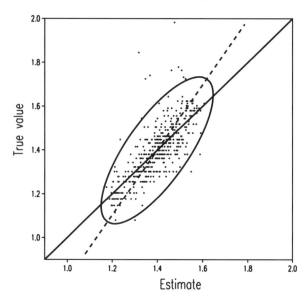

Figure 8.24 Scatter diagram of the true $\log_{10} K$ for Broom's Barn Farm plotted against the punctually kriged estimates. The ellipse is a probability contour, the dashed line is its longer diameter, and the solid diagonal line is the regression of $Z(\mathbf{x}_0)$ on $\hat{Z}(\mathbf{x}_0)$.

$45°$. The variance of the estimates, 0.009 79 on the abscissa, is less than that of the true values, 0.018 00 on the ordinate. In other words, kriging has lost variance; kriging smooths. It underestimates the larger values and overestimates the smaller ones, as in the simpler forms of regression.

9

Cross-correlation, coregionalization and cokriging

9.1 INTRODUCTION

In this chapter we develop the ideas of spatial correlation in individual variables for use in situations in which two or more environmental variables interest us simultaneously. We shall assume that each variable individually can be treated as if it were random, so that all of the statistical theory and techniques of Chapters 4–7 apply. We shall use the data from two surveys to illustrate the development. One set comprises the exchangeable potassium (K), available phosphorus (P) and yield of barley in the topsoil of a 6.4 ha field in southeast England (CEDAR Farm, Centre for Dairy Research); and the other comprises the concentrations of potentially toxic trace metals in the Swiss Jura. Table 9.1 summarizes the data for the Farm, and Table 9.5 that of the Jura.

There are now two additional features of the variation to consider. One comprises the relations between variables, regardless of space, as expressed in the ordinary product-moment correlation, r, of equation (2.11). The correlation matrix for CEDAR Farm is given in Table 9.2. Evidently K, P and yield are related, though not strongly. In the Swiss Jura the correlations among the trace metals in the soil are stronger (Table 9.6), and we might wish to consider them all together in assessing the risk of pollution. The other feature concerns the spatial aspects of this correlation: one variable may be spatially related with another in the sense that its values at places are correlated with the values of the other variable. For example, the potassium and phosphorus in the soil at CEDAR Farm might be spatially correlated with the crop yield, and the cadmium with the zinc in the Jura. In these circumstances we might be able to take advantage of the correlation and the information contained in the several variables to predict any one of them.

Table 9.1 Summary statistics of K, P and yield of barley at CEDAR Farm based on a sample of $N = 160$.

	K $(mg\ kg^{-1})$	P $(mg\ kg^{-1})$	Yield $(t\ ha^{-1})$
Minimum	101.0	16.8	1.28
Maximum	243.0	89.0	4.43
Mean	155.7	49.4	3.03
Median	151.0	50.9	3.11
St. dev.	28.7	14.4	0.50
Variance	825.65	206.84	0.249
Skewness	0.91	−0.24	−0.72

Table 9.2 Correlation matrix for K, P and yield at CEDAR Farm with 158 degrees of freedom.

K	1		
P	0.585	1	
Yield	−0.329	−0.395	1
	K	P	Yield

We formalize these ideas under the general heading of *coregionalization*. We start by considering two regionalized variables, $Z_u(\mathbf{x})$ and $Z_v(\mathbf{x})$, which we shall denote u and v, both obeying the intrinsic hypothesis. Thus for variable u we have, from Chapter 4,

$$E[Z_u(\mathbf{x}) - Z_u(\mathbf{x} + \mathbf{h})] = 0.$$

The variable will also have a variogram, specifically an *autovariogram*:

$$\gamma_{uu}(\mathbf{h}) = \tfrac{1}{2}E[\{Z_u(\mathbf{x}) - Z_u(\mathbf{x} + \mathbf{h})\}^2]. \tag{9.1}$$

The reason for the double uu will become apparent presently. Similarly for v, the expected differences are 0, and its autovariogram is $\gamma_{vv}(\mathbf{h})$, consisting of the expected squared differences in v.

The two variables will also have a *cross-variogram*, $\gamma_{uv}(\mathbf{h})$, defined as

$$\gamma_{uv}(\mathbf{h}) = \tfrac{1}{2}E[\{Z_u(\mathbf{x}) - Z_u(\mathbf{x} + \mathbf{h})\}\{Z_v(\mathbf{x}) - Z_v(\mathbf{x} + \mathbf{h})\}]. \tag{9.2}$$

This function describes the way in which u is related spatially to v.

If both variables are second-order stationary with means μ_u and μ_v, then both will have covariance functions:

$$C_{uu}(\mathbf{h}) = E[\{Z_u(\mathbf{x}) - \mu_u\}\{Z_u(\mathbf{x} + \mathbf{h}) - \mu_u\}] \tag{9.3}$$

and analogously for $C_{vv}(\mathbf{h})$. They will also have a cross-covariance function:

$$C_{uv}(\mathbf{h}) = \mathrm{E}[\{Z_u(\mathbf{x}) - \mu_u\}\{Z_v(\mathbf{x} + \mathbf{h}) - \mu_v\}]. \tag{9.4}$$

As in the univariate case, there is a spatial cross-correlation coefficient, $\rho_{uv}(\mathbf{h})$, which is given by

$$\rho_{uv}(\mathbf{h}) = \frac{C_{uv}(\mathbf{h})}{\sqrt{C_{uu}(\mathbf{0})C_{vv}(\mathbf{0})}}. \tag{9.5}$$

Equation (9.5) is the extension of the ordinary Pearson product-moment correlation coefficient (Chapter 2) into the spatial domain for $Z_u(\mathbf{x})$ and $Z_v(\mathbf{x} + \mathbf{h})$. When $\mathbf{h} = \mathbf{0}$ it is the Pearson coefficient. Note, however, that $\rho_{uv}(\mathbf{0}) = 0$, i.e. no linear correlation in the usual sense, does not mean no correlation at lag distances greater than zero.

Equation (9.4) contains another new feature, namely asymmetry, for in general

$$\mathrm{E}[\{Z_u(\mathbf{x}) - \mu_u\}\{Z_v(\mathbf{x} + \mathbf{h}) - \mu_v\}] \neq \mathrm{E}[\{Z_v(\mathbf{x}) - \mu_v\}\{Z_u(\mathbf{x} + \mathbf{h}) - \mu_u\}]. \tag{9.6}$$

In words, the cross-covariance between u and v in one direction is in general different from that in the opposite direction; the function is asymmetric:

$$C_{uv}(\mathbf{h}) \neq C_{uv}(-\mathbf{h}) \quad \text{or equivalently} \quad C_{uv}(\mathbf{h}) \neq C_{vu}(\mathbf{h}),$$

since

$$C_{uv}(\mathbf{h}) = C_{vu}(-\mathbf{h}).$$

Asymmetry in time is common. The temperature of the air during the day reaches its maximum after the sun has reached its zenith and its minimum occurs after midnight, and the air's mean daily temperature has maxima and minima after the solstices. There is a delay between the elevation of the sun and the temperature of the air. Analogous asymmetry in one dimension in space is easy to envisage. The topsoil might be related asymmetrically to the subsoil on a slope as a result of soil creep, and irrigation by periodic flooding from the same end of a field might redistribute salts differentially down the profile. However, unless the evidence for asymmetry is strong or there is some physical rationale for spatial asymmetry, one might treat differences in estimates (equation (9.10) below) as sampling effects and proceed as though the cross-correlation is symmetric.

The cross-variogram and the cross-covariance function (if it exists) are related, and as in the univariate case the variogram can be obtained from the covariance function by extension of equation (4.4), as follows:

$$\gamma_{uv}(\mathbf{h}) = C_{uv}(\mathbf{0}) - \tfrac{1}{2}\{C_{uv}(\mathbf{h}) + C_{uv}(-\mathbf{h})\}. \tag{9.7}$$

However, this conversion does not retain all of the information, as we can see by splitting the cross-covariance into an even and an odd term:

$$C_{uv}(\mathbf{h}) = \tfrac{1}{2}\{C_{uv}(+\mathbf{h}) + C_{uv}(-\mathbf{h})\} + \tfrac{1}{2}\{C_{uv}(+\mathbf{h}) - C_{uv}(-\mathbf{h})\}. \qquad (9.8)$$

The odd term, the second term on the right-hand side of equation (9.8), does not appear in equation (9.7). Unlike the cross-covariance, therefore, the cross-variogram is an even function, i.e. it is symmetric:

$$\gamma_{uv}(\mathbf{h}) = \gamma_{vu}(\mathbf{h}) \qquad \text{for all } \mathbf{h}.$$

The cross-variogram cannot express asymmetry, and it should not be used where asymmetry is thought to be significant.

Another way of expressing the spatial relations between the two variables is by the codispersion coefficient. For a lag $\mathbf{h}$, this is

$$\nu_{uv}(\mathbf{h}) = \frac{\gamma_{uv}(\mathbf{h})}{\sqrt{\gamma_{uu}(\mathbf{h})\gamma_{vv}(\mathbf{h})}}. \qquad (9.9)$$

This coefficient may be thought of as the correlation between the spatial differences of u and v. Its merit is that it is symmetric, and so its estimate might be preferred to the cross-correlogram (9.5) for describing the cross-correlation. For second-order stationarity, $\nu_{uv}(\mathbf{h})$ approaches $\rho_{uv}(\mathbf{0})$ as $|\mathbf{h}|$ approaches infinity.

9.2 ESTIMATING AND MODELLING THE CROSS-CORRELATION

Provided there are sites where both u and v have been measured, $\gamma_{uv}(\mathbf{h})$ can be estimated in a way similar to that for autosemivariances by

$$\hat{\gamma}_{uv}(\mathbf{h}) = \frac{1}{2m(\mathbf{h})} \sum_{i=1}^{m(\mathbf{h})} \{z_u(\mathbf{x}_i) - z_u(\mathbf{x}_i + \mathbf{h})\}\{z_v(\mathbf{x}_i) - z_v(\mathbf{x}_i + \mathbf{h})\}. \qquad (9.10)$$

The result is an experimental cross-variogram for u and v.

The cross-variogram can be modelled in the same way as that of autovariograms, and the same restricted set of functions is available. To describe the coregionalization there is an added condition. Any linear combination of the variables is itself a regionalized variable, and its variance must be positive or zero: it may not be negative. This is ensured as follows.

We adopt what is called the linear model of coregionalization. In it we assume that each variable $Z_u(\mathbf{x})$ is a linear sum of orthogonal, i.e. independent, random variables $Y_j^k(\mathbf{x})$, each with mean 0 and variance 1, and in which the superscript k is simply an index, not a power:

$$Z_u(\mathbf{x}) = \sum_{k=1}^{K} \sum_{j=1}^{2} a_{uj}^k Y_j^k(\mathbf{x}) + \mu_u. \qquad (9.11)$$

In this expression

$$E[Z_u(\mathbf{x})] = \mu_u,$$

$$E[Y_j^k(\mathbf{x})] = 0 \qquad \text{for all } k \text{ and } j,$$

and

$$\tfrac{1}{2}E[\{Y_j^k(\mathbf{x}) - Y_j^k(\mathbf{x} + \mathbf{h})\}\{Y_{j'}^{k'}(\mathbf{x}) - Y_{j'}^{k'}(\mathbf{x} + \mathbf{h})\}]$$

$$= \begin{cases} g_k(\mathbf{h}) > 0, & \text{for } k = k' \text{ and } j = j', \\ 0, & \text{otherwise.} \end{cases}$$

Then the variogram for any pair of variables u and v is

$$\gamma_{uv}(\mathbf{h}) = \sum_{k=1}^{K} \sum_{j=1}^{2} a_{uj}^k a_{vj}^k g_k(\mathbf{h}). \tag{9.12}$$

We can replace the products in the second summation by b_{uv}^k to obtain

$$\gamma_{uv}(\mathbf{h}) = \sum_{k=1}^{K} b_{uv}^k g_k(\mathbf{h}). \tag{9.13}$$

These b_{uv}^k are the variances and covariances, i.e. nugget and sill variances, for the independent components if they are bounded. The result might look like the set of spherical-plus-nugget functions in Figure 9.3 below. The intercepts are the three nugget variances, b^1, and the differences between these and the maxima are the sills of the correlated variances, b^2. For unbounded variograms the b_{uv}^k are the nugget variances and gradients. The coefficients $b_{uv}^k = b_{vu}^k$ for all k, and for each k the matrix of coefficients

$$\begin{bmatrix} b_{uu}^k & b_{uv}^k \\ b_{vu}^k & b_{vv}^k \end{bmatrix}$$

must be positive definite. Since the matrix is symmetric, it is sufficient that $b_{uu}^k \geqslant 0$ and $b_{vv}^k \geqslant 0$ and that its determinant is positive or zero:

$$|b_{uv}^k| = |b_{vu}^k| \leqslant \sqrt{b_{uu}^k b_{vv}^k}.$$

This is Schwarz's inequality.

For V coregionalized variables the full matrix of coefficients, $[b_{ij}]$, will be of order V, and its determinant and all its principal minors must be positive or zero.

Schwarz's inequality has the following consequences for each pair of variables:

1. Every basic structure, $g_k(\mathbf{h})$, represented in a cross-variogram must also appear in the two autovariograms, i.e. $b_{uu}^k \neq 0$ and $b_{vv}^k \neq 0$ if $b_{uv}^k \neq 0$.

   As a corollary, if a basic structure $g_k(\mathbf{h})$ is absent from either auto-variogram, then it may not be included in the cross-variogram.

2. The reverse is not so: b_{uv}^k may be zero when $b_{uu}^k > 0$, and structures may be present in the autovariograms without their appearing in the cross-variogram.

In practice, fitting an optimal model to the coregionalization with these constraints seems formidable. Nevertheless, Goulard and Voltz (1992) have provided an algorithm that converges swiftly. One chooses a suitable combination of basic variogram functions, say nugget plus spherical, and for the autocorrelated function(s) one provides the distance parameters. These can be approximated in advance by fitting models independently to the experimental variograms. Starting with reasonable values for the coefficients, b_{uv}^k, the computer fits the model and then iterates to minimize the residual sum of squares, checking at each step that the solution is CNSD.

As a check on the validity of a model of coregionalization one can plot the cross experimental variogram for any pair of variables and the model for them plus the limiting values that would hold if correlation were perfect. This last gives what Wackernagel (1995) calls the 'hull of perfect correlation', and for any pair of variables u and v it is obtained from the coefficients b_{uu}^k and b_{vv}^k by

$$\text{hull}[\gamma_{uv}(\mathbf{h})] = \pm \sum_{k=1}^{K} \sqrt{b_{uu}^k b_{vv}^k} g_k(\mathbf{h}). \tag{9.14}$$

The proximity of the line of the model to the experimental points shows the goodness of fit, as before (Chapter 6). The line must also lie within the hull to be acceptable. But perhaps most revealing is the proximity of the cross-variogram to the hull. If the two are close then the cross-correlation is strong. If, in contrast, the cross-variogram lies far from the bounds then the correlation is weak. This feature may be appreciated by examining Figure 9.3, and we shall discuss it in the first example below.

9.2.1 Intrinsic coregionalization

In general, the ratios of the coefficients to one another vary from one basic function to another. In Figure 9.1(a), for example, we have a simple nugget-plus-spherical variogram,

$$y(h) = 2 + 8\,\text{sph}(1.7),$$

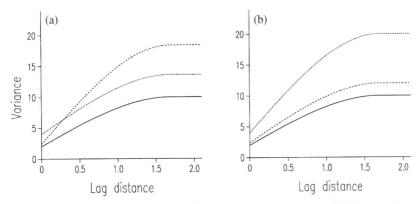

Figure 9.1   Spherical variograms with constant range, 1.7: (a) of differing shapes and differing nugget:sill ratios in the general case; (b) of constant nugget:sill ratios in the intrinsic case. See text for further explanation.

shown as a solid line. If we multiply the nugget, 2, by 1.2 and the spherical component by 2 we obtain the dashed line, which has a different shape from the solid line. If our two multipliers are 2 and 1.2 then we obtain the dotted line, which is of a similar shape to the first, but with a different nugget:sill ratio. The range is the same, but the proportions of nugget to sill are all different.

It sometimes happens, however, that all the auto- and cross-variograms are proportional to a single variogram function, so that in terms of equation (9.13) all the coefficients b_{uv}^k are the same for all k for each combination of u and v, thus:

$$\gamma_{uv}(\mathbf{h}) = \sum_{k=1}^{K} b_{uv} g_k(\mathbf{h}), \qquad (9.15)$$

in which we replace the b_{uv}^k, $k = 1, 2, \ldots, K$, by the single coefficient b_{uv}. They are simply multiples of one another with the same basic shape. As an example, Figure 9.1(b) shows the basic spherical variogram. If we multiply the two original components in turn by 1.2 and 2, representing b_{uu}, b_{vv} and b_{uv}, then we obtain the two additional variograms, represented by the dashed and dotted lines, respectively. These are the same apart from the vertical scale.

Where the variables are second-order stationary,

$$\gamma_{uv}(\mathbf{h}) = C_{uv}(0)g(\mathbf{h}), \qquad (9.16)$$

with $g(\mathbf{h}) \to 1$ as $|\mathbf{h}| \to \infty$. Spatial cross-correlation of this kind is said to be *intrinsic*. The term is somewhat unfortunate in that this usage of 'intrinsic' differs from that in the 'intrinsic hypothesis'.

Where spatial correlation is intrinsic the codispersion coefficient $v_{uv}(\mathbf{h})$ remains constant for all $\mathbf{h}$, i.e.

$$v_{uv}(\mathbf{h}) = \frac{b_{uv}}{\sqrt{b_{uu}b_{vv}}} = \frac{C_{uv}(0)}{\sqrt{C_{uu}(0)C_{vv}(0)}} = v_{uv}(0).$$

9.3   EXAMPLE: CEDAR FARM

We illustrate the procedure and some of the features of coregionalization using the survey data of a field on CEDAR Farm in south-east England. They derive from an original study of precision farming by Dr Z. L. Frogbrook, who kindly provided them and to whom we are grateful. The field covers approximately 6.4 ha of fairly flat land on clay and is cultivated to produce cereals. Its topsoil (0–15 cm) was sampled at 160 places on 5 m × 2 m supports at 20 m intervals on a square grid. The yield of barley was measured on the same supports in 1998. The principal plant nutrients, exchangeable potassium (K) and available phosphorus (P), were measured. The data are summarized in Table 9.1. Potassium and yield are somewhat skewed, but not so seriously as to warrant transformation. The three variables are correlated, though not strongly, as Table 9.2 shows.

By applying equation (9.10) and treating the variation as isotropic, we obtain the experimental auto- and cross-variograms. The experimental variograms have simple forms. Figure 9.2 shows an example; it is the autovariogram of K with a spherical model fitted to it by weighted least squares, as described in Chapter 6. The model's coefficients are listed in Table 9.3. The other experimental variograms appear in Figure 9.3 as the point symbols. Four of them, namely the autovariogram of yield (Figure 9.3(e)), and the three cross-variograms (Figures 9.3(b), 9.3(d) and 9.3(f)) are evidently bounded, and again the spherical model fitted them well. The coefficients of fitting the model independently are listed in Table 9.3. The autovariogram of P does not reach a bound within the field.

The five bounded variograms have approximately the same range, and so we can reasonably fit the linear model of coregionalization with two basic components, $g_1(0)$, i.e. nugget, and $g_2(|\mathbf{h}|)$. We set the range of $g_2(|\mathbf{h}|)$ to 144 m, the average of the five bounded variogram models. The resulting coefficients b_{uv}^k are listed in Table 9.4, and the solid lines in Figure 9.3 are those of the linear model of coregionalization. We can see by comparing Figure 9.2 with Figure 9.3(a) for K that the model of coregionalization fits somewhat less well than the model fitted to K alone, and we can imagine that we could fit the other variograms better if we treated them individually. Nevertheless, the fit is generally good, and even for P the model fits fairly well over most of the working lag distance. The dashed lines in Figures 9.3(b), 9.3(d) and 9.3(f) are the hulls of perfect correlation for the cross-variograms. In all three the model is some way from

Table 9.3 Coefficients of spherical model fitted independently to auto- and cross-variograms of K, P and barley yield at CEDAR Farm. The parameters c_0, c and a are the nugget, sill of the correlated variance and the range, respectively.

	c_0	c	a (m)
K	63.7	855.4	121.9
P	25.2	—	—
Yield	0.121	0.167	151.0
K × P	−20.48	294.6	150.2
K × yield	−1.51	7.94	130.1
P × yield	−0.550	4.548	168.5

Table 9.4 Coefficients b_{uv}^k of the linear coregionalization with nugget plus spherical for K, P and yield at CEDAR Farm.

		K	P	Yield
Nugget, b_{uv}^1				
	K	157.28		
	P	−21.34	2.896 0	
	Yield	0.868 3	−0.113 05	0.150 27
[6pt] Correlated variance, b_{uv}^2				
	K	785.15		
	P	291.39	218.199 6	
	Yield	−7.423 9	−3.514 9	0.130 27

the hulls, showing that the spatial cross-correlations are at best moderate and like the ordinary simple correlations (Table 9.2).

Finally, we comment on the practical meaning of the coregionalization. The correlation between K and P is positive, and all the cross-semivariances are positive. The two variables characterize the nutrient status of the soil. The correlations between yield and K and P are all negative, and this might come as a surprise. Farmers and their advisers have been used to thinking that more K and P in the soil would result in greater yield. Now that yield can be recorded automatically at harvest, they are discovering that large yields deplete the soil locally and that they should fertilize differentially to maintain sufficient K and P over the whole of each field. Large concentrations of K and P indicate small off-take of these nutrients by the crop and thus smaller yields. Webster, in Lake *et al.* (1997, pp. 74–77), has shown another example from the same district.

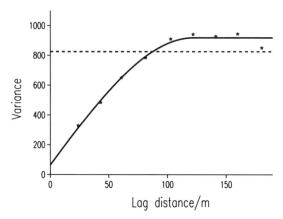

Figure 9.2 Variogram of potassium at CEDAR Farm with experimental values plotted as point symbols and the spherical model shown as the solid line.

9.4 COKRIGING

Having learned how to model the coregionalization, we can use our knowledge of the spatial relations between two or more variables to predict their values by cokriging. Typically the aim is to estimate just one variable, which we may regard as the principal or target variable, at a point x_0 or in a block B from data on it plus those of one or more other variables, which we regard as subsidiary variables. Cokriging is simply an extension of autokriging in that it takes into account additional correlated information in the subsidiary variables. It appears more complex because the additional variables increase the notation.

Let there be V variables, $l = 1, 2, \ldots, V$, and let us denote the one we wish to predict as u; this will usually have been less densely sampled than the others. In ordinary cokriging we form the linear sum

$$\hat{Z}_u(B) = \sum_{l=1}^{V} \sum_{i=1}^{n_l} \lambda_{il} z_l(\mathbf{x}_i),$$ (9.17)

where the subscript l refers to the variables, of which there are V, and the subscript i refers to the sites, of which there are n_l where the variable l has been measured. The λ_{il} are weights, satisfying

$$\sum_{i=1}^{n_l} \lambda_{il} = \begin{cases} 1, & l = u, \\ 0, & l \neq u. \end{cases}$$ (9.18)

These are the non-bias conditions, and subject to them the estimation variance of $\hat{Z}_u(B)$ for a block, B, is minimized by solving the kriging sys-

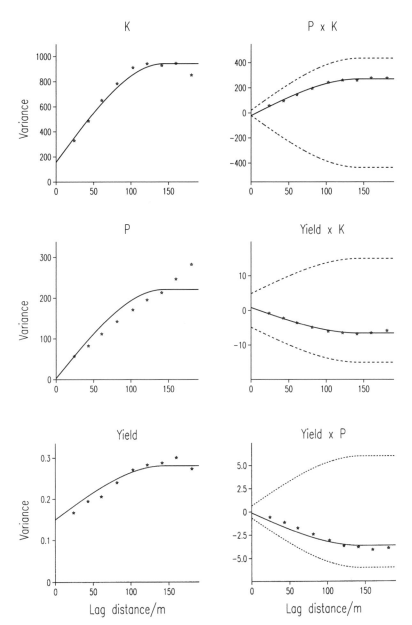

Figure 9.3   Autovariograms (left) and cross-variograms (right) of K, P and barley yield at CEDAR Farm. The experimental values are plotted as points and the solid lines are of the model of coregionalization. The dashed lines in the right-hand graphs are the hulls of perfect correlation.

tem, which, in full, is

$$\sum_{l=1}^{V}\sum_{i=1}^{n_l}\lambda_{il}y_{lv}(\mathbf{x}_i,\mathbf{x}_j) + \psi_v = \overline{y}_{uv}(\mathbf{x}_j,B),$$

$$\sum_{i=1}^{n_l}\lambda_{il} = \begin{cases} 1, & l = u, \\ 0, & l \neq u. \end{cases} \tag{9.19}$$

for all $v = 1, 2, \ldots, V$ and all $j = 1, 2, \ldots, n_v$. The quantity $y_{lv}(\mathbf{x}_i, \mathbf{x}_j)$ is the (cross-)semivariance between variables l and v at sites i and j, separated by the vector $\mathbf{x}_i - \mathbf{x}_j$; $\overline{y}_{uv}(\mathbf{x}_j, B)$ is the average (cross-)semivariance between a site j and the block B, and ψ_v is the Lagrange multiplier for the vth variable. We print 'cross' in parentheses because if $l = v$ or $u = v$ the semivariances are the autosemivariances. This set of equations is the extension of the autokriging system, equations (8.11).

Solving equations (9.19) gives the weights, λ, which are inserted into equation (9.17) to estimate $Z_u(B)$, and the estimation variance, the cokriging variance, is obtained from

$$\sigma_u^2(B) = \sum_{l=1}^{V}\sum_{j=1}^{n_l}\lambda_{jl}\overline{y}_{ul}(\mathbf{x}_j,B) + \psi_u - \overline{y}_{uu}(B,B), \tag{9.20}$$

where $\overline{y}_{uu}(B,B)$ is the integral of $y_{uu}(\mathbf{h})$ over B, i.e. the within-block variance of u.

The equations can be represented in matrix form. For simplicity consider two variables, u and v, only. The matrices are easily extended to more. Let Γ_{uv} denote a matrix of semivariances (including cross-semivariances where $u \neq v$) between sampling points in a neighbourhood. Let there be n_u places at which variable u was measured and n_v where v was measured. The order of the matrix is $n_u \times n_v$:

$$\Gamma_{uv} = \begin{bmatrix} y_{uv}(\mathbf{x}_1,\mathbf{x}_1) & y_{uv}(\mathbf{x}_1,\mathbf{x}_2) & \cdots & y_{uv}(\mathbf{x}_1,\mathbf{x}_v) \\ y_{uv}(\mathbf{x}_2,\mathbf{x}_1) & y_{uv}(\mathbf{x}_2,\mathbf{x}_2) & \cdots & y_{uv}(\mathbf{x}_2,\mathbf{x}_v) \\ \vdots & \vdots & \cdots & \vdots \\ y_{uv}(\mathbf{x}_{n_u},\mathbf{x}_1) & y_{uv}(\mathbf{x}_{n_u},\mathbf{x}_2) & \cdots & y_{uv}(\mathbf{x}_{n_u},\mathbf{x}_{n_v}) \end{bmatrix}.$$

We denote by $\mathbf{b}_{uu}$ and by $\mathbf{b}_{uv}$ the vectors of autosemivariances for variable u and cross-semivariances:

$$\mathbf{b}_{uu} = \begin{bmatrix} \overline{y}_{uu}(\mathbf{x}_1,B) \\ \overline{y}_{uu}(\mathbf{x}_2,B) \\ \vdots \\ \overline{y}_{uu}(\mathbf{x}_{n_u},B) \end{bmatrix}, \quad \mathbf{b}_{uv} = \begin{bmatrix} \overline{y}_{uv}(\mathbf{x}_1,B) \\ \overline{y}_{uv}(\mathbf{x}_2,B) \\ \vdots \\ \overline{y}_{uv}(\mathbf{x}_{n_v},B) \end{bmatrix}.$$

The matrix equation is then

$$
\begin{bmatrix}
 & & 10 \\
\Gamma_{uu} & \Gamma_{uv} & 10 \\
 & & \vdots \\
 & & 10 \\
 & & 01 \\
\Gamma_{vu} & \Gamma_{vv} & 01 \\
 & & \vdots \\
 & & 01 \\
11\ldots1 & 00\ldots0 & 00 \\
00\ldots0 & 11\ldots1 & 00
\end{bmatrix}
\cdot
\begin{bmatrix}
\lambda_{1u} \\
\lambda_{2u} \\
\vdots \\
\lambda_{n_u u} \\
\lambda_{1v} \\
\lambda_{2v} \\
\vdots \\
\lambda_{n_v v} \\
\psi_u \\
\psi_v
\end{bmatrix}
=
\begin{bmatrix}
\mathbf{b}_{uu} \\
\\
\mathbf{b}_{uv} \\
\\
1 \\
0
\end{bmatrix}.
$$

If we denote the augmented matrix of Γs by $\mathbf{G}$, the vector of weights and Lagrange multipliers by $\boldsymbol{\lambda}$, and the right-hand side vector by $\mathbf{b}$, then we can write the solution of the equation succinctly as

$$\boldsymbol{\lambda} = \mathbf{G}^{-1}\mathbf{b}. \tag{9.21}$$

The cokriging (prediction) variance is given by

$$\hat{\sigma}_u^2(B) = \mathbf{b}^\mathsf{T}\boldsymbol{\lambda} - \overline{\gamma}_{uu}(B, B). \tag{9.22}$$

As in autokriging, the block B may be of any reasonable size and shape, and it may be reduced to a point, $\mathbf{x}_0$, having the same dimensions as the support on which the data were obtained. In these circumstances the averages $\overline{\gamma}_{uv}(\mathbf{x}_j, B)$ become $\gamma_{uv}(\mathbf{x}_j, \mathbf{x}_0)$, and $\overline{\gamma}_{uu}(B, B)$ is zero and hence disappears, thus:

$$
\mathbf{b}_{uu} =
\begin{bmatrix}
\gamma_{uu}(\mathbf{x}_1, \mathbf{x}_0) \\
\gamma_{uu}(\mathbf{x}_2, \mathbf{x}_0) \\
\vdots \\
\gamma_{uu}(\mathbf{x}_{n_u}, \mathbf{x}_0)
\end{bmatrix},
\qquad
\mathbf{b}_{uv} =
\begin{bmatrix}
\gamma_{uv}(\mathbf{x}_1, \mathbf{x}_0) \\
\gamma_{uv}(\mathbf{x}_2, \mathbf{x}_0) \\
\vdots \\
\gamma_{uv}(\mathbf{x}_{n_v}, \mathbf{x}_0)
\end{bmatrix},
$$

and

$$\hat{\sigma}_u^2(\mathbf{x}_0) = \mathbf{b}^\mathsf{T}\boldsymbol{\lambda}. \tag{9.23}$$

Myers (1982) presents the equations for cokriging somewhat differently and comprehensively.

9.4.1 Is cokriging worth the trouble?

Cokriging is more complex than autokriging, and the practitioner can and should ask whether the extra complexity improves the results: are the estimates better in any sense?

We distinguish two situations. First consider the *undersampled* case. By undersampling we mean that the variable to be estimated, the primary variable u in the kriging equations, is sampled less intensely than the others, usually at a subset of the sampling points. In this case the spatial correlation in the other variables and their relation to u add information that is lacking in that of u alone. As a result cokriging increases the precision, i.e. it reduces the estimation variance. By how much depends on the degree of undersampling. In general, the smaller the sampling intensity of u in relation to that of the other(s) the greater is the benefit of cokriging. We illustrate this below.

In the *fully sampled* case, all variables are recorded at all sampling points. Here the principal advantage is *coherence*. Kriging is coherent when the kriged estimate of the sum of a set of variables, say $\hat{S}$, equals the sum of their individually kriged estimates:

$$\hat{S}(B) = \sum_{i=1}^{n_l} \lambda_i \sum_{l=1}^{V} z_l(\mathbf{x}_i) = \sum_{k=1}^{V} \sum_{l=1}^{V} \sum_{i=1}^{n_l} \lambda_{il} z_l(\mathbf{x}_i). \qquad (9.24)$$

Cokriging ensures coherence. Otherwise the equality depends on the nature of the coregionalization.

As an example, consider estimating the thickness of a soil horizon. A field surveyor might have recorded the depths from the surface to the top and bottom of the horizon in question at the sampling points. At each point the thickness is simply the difference between the two. We could krige the thickness to estimate it at unrecorded positions. Alternatively, we could krige the depths to the top and the bottom of the horizon and compute their differences. If we were to do that for each variable independently then we should find that in general the differences between the kriged estimates were not the same as the kriged differences, i.e. the kriged thickness. If, however, we were to cokrige the depths to top and bottom then the differences between the kriged estimates would equal the kriged thickness.

Where the variables are intrinsically coregionalized, i.e. all the variograms are related linearly to a single basic model, autokriging of any variable gives the same result as cokriging. The spatial information for the one variable is all there is in the data, there is no more in that of the others, and there is no merit in the more complex procedure.

Where the variogram of the primary variable, u, is linearly related to the cross-variogram(s), autokriging u again gives results identical with cokriging. The cross-correlation adds nothing.

In other situations the results are in general different. However, with full sampling the differences are likely to be small, and experience suggests that the differences are usually so small that they can be ignored unless coherence is essential.

9.4.2   Example of benefits of cokriging

We can see something of the benefits of cokriging by following the same logic as in Chapter 8, where we calculated the kriging variances for various sample spacings from a model variogram. We placed sampling points on regular grids, and we computed the variances at the centres of the grid cells where for punctual kriging they were greatest. For block kriging the maxima can occur when the target block is centred on a grid node, and so we calculated the variance in those positions also. We displayed the results as curves of maximum kriging variance against grid spacing (Figure 8.23).

For cokriging we have two or more variables. We can choose the primary grid for the undersampled target variable, u, in the same way as for autokriging. We can then superimpose denser grids for the subsidiary variables. With the points at the nodes of these two grids we can set up and solve the cokriging equations. The maximum kriging variances are no longer necessarily at the centres of grids cells or centred over grid nodes, and so their positions must be found by searching. As the density of the subsidiary grid is increased so the kriging variance of the target variable should decrease, and doing the calculations as above and plotting the results will show just how beneficial cokriging is at various scales. McBratney and Webster (1983) describe the procedure in detail.

We illustrate the approach with the coregionalization at CEDAR Farm. As we mentioned above, measuring nutrients in the soil is expensive in relation to the benefits to be gained from knowing the concentrations. Arable farmers in Britain can afford to sample their soil at a density of approximately one per hectare; not more. Yet they would like to know the nutrient concentration at a much finer resolution to vary their application of fertilizers. Automatic recording of yield as the grain passes through the harvester is now quite feasible and produces abundant dense data. So if the relation between yield and nutrient status is sufficiently strong the farmer might use the dense data on yield to improve his prediction of nutrient concentration. So let us see to what extent we might use this approach in the situation at this farm.

We suppose that available phosphorus (P) is the target variable and that we shall use yield as the subsidiary variable. We take the parameters for the cokriging from the coregionalization model (Table 9.4). We have chosen intervals for the primary grid from near 0 to 400 m. We have imposed subsidiary grids with intervals of 1/2, 1/3, 1/4, and 1/5 of the primary grid, giving sampling ratios of 4, 9, 16, and 25. The smallest intervals are impracticable because the cutter bar of a modern harvester is typically 4 m wide on British farms, but we include them to complete the picture and for theoretical interest. We have solved the kriging systems for punctual kriging and also computed the kriging variances for blocks 24 m × 24 m. We choose this size because the standard farm machinery spreads fertilizer in bands this wide.

The results are plotted as graphs of maximum kriging variance against sample spacing in Figure 9.4. In each graph the uppermost solid curve is for autokriging and the ones beneath it are in order from top to bottom for cokriging with the subsidiary grid interval 1/2, 1/3, 1/4 and 1/5 of that of the primary grid.

The upper pair of graphs, Figure 9.4(a) for punctual kriging and Figure 9.4(b) for block kriging, show that with the actual model of coregionalization for this field the reductions in kriging variance from adding yield in the kriging equations to predict P are modest. The reason is that the correlation between the two is itself modest. If the cost of installing a recorder to measure yield and handling the data is much less than that of analysing the soil for P then it might be worth the trouble, but in any event the farmer cannot expect large gains in precision or to save much in soil sampling and analysis.

In passing, we note that the block-kriging variance is less than the variance of punctual kriging with the same sampling configuration by an amount approximately equal to the within-block variance of P.

The outlook might be rosier with stronger association between target and subsidiary variables, and to illustrate this we have repeated the exercise using a model of perfect correlation, the lower part of the hull in Figure 9.3(f). Figure 9.4(c)–(d) displays the results. Now large differences emerge as the density of the subsidiary grid increases. With a sampling ratio of 25 for the subsidiary variable we can reduce the maximum kriging variance to one-third of that from autokriging. The farmer could increase the ratio further and gain even bigger benefits in such circumstances.

9.5   PRINCIPAL COMPONENTS OF COREGIONALIZATION MATRICES

The full coregionalization model of V variables has $V \times (V + 1)/2$ variograms, and if each has K basic functions then there are $K \times V$ coefficients. We have already seen the set of auto- and cross-variograms for CEDAR Farm, with three variables and $K = 2$ basic functions (Table 9.4 and Figure 9.3). The correlations are moderate, and though we have used the relation between P and yield to illustrate cokriging there is not a great deal of interest in exploring them further. Where the correlations are stronger, however, it may be worth analysing the coregionalization matrices to see how the correlation varies with scale. To illustrate this we turn to an original investigation of heavy metals in the soil in the Swiss Jura by Atteia *et al.* (1994) and Webster *et al.* (1994).

Some 14.5 km^2 near La Chaux-de-Fonds in the Jura were surveyed to determine the concentrations of seven potentially toxic metals, namely

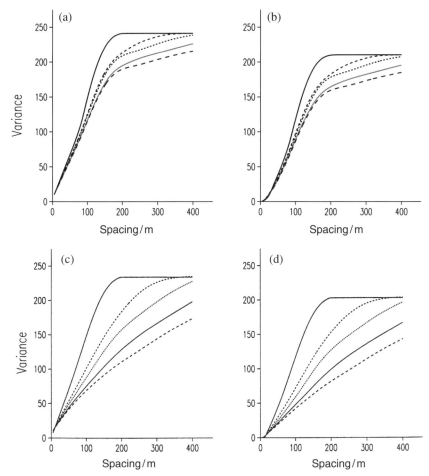

Figure 9.4 Graphs of maximum kriging variance of phosphorus (P) against sample spacings on a primary grid with denser observations of yield on subsidiary grids. (a) Punctual kriging of P using the fitted model of coregionalization (Table 9.4). (b) Kriging of 24 m × 24 m blocks with the same model. (c) Punctual kriging of P using a model of perfect correlation (Figure 9.3(f)). (d) Block kriging (24 m × 24 m blocks) with the perfect model. In each graph the uppermost curve is that for autokriging and the ones below are in order for spacings on the subsidiary grid of 1/2, 1/3, 1/4 and 1/5 of those on the primary grid.

cadmium (Cd), cobalt (Co), chromium (Cr), copper (Cu), nickel (Ni), lead (Pb) and zinc (Zn), in the topsoil. Soil was removed with a cylindrical corer of 5 cm diameter to a depth of 25 cm, which therefore defined the support of the sample. Cores were taken at 214 intersections of a 250 m grid plus an additional 152 points arranged in nests around 38 of the grid nodes. The 'total' metal was extracted from each sample by strong acid and mea-

Table 9.5 Summary statistics for heavy metals at La Chaux-de-Fonds on original scales and with Cd, Cu, Pb and Zn transformed to their common logarithms.

	Cd	Co	Cr	Cu	Ni	Pb	Zn
Minimum	0.14	1.55	3.32	3.55	1.98	18.7	25.0
Maximum	5.13	20.6	70.0	242	53.2	382.0	338.0
Mean	1.31	9.45	35.2	24.6	20.2	57.0	78.5
Median	1.11	9.82	34.8	17.4	20.8	46.8	74.0
Variance	0.7598	12.56	118.3	638.8	67.91	1527.9	1147.0
St. dev.	0.87	3.54	10.9	25.3	8.24	39.1	33.9
Skewness	1.43	−0.20	0.34	3.87	0.17	4.22	2.74
Kurtosis	2.44	−0.60	0.33	18.1	0.31	23.6	12.9
Logarithms							
Mean	0.022			1.26		1.70	1.86
Variance	0.0868			0.1046		0.0414	0.0306
St. dev.	0.29			0.34		0.21	0.18
Skewness	−0.30			0.51		1.10	0.07
Kurtosis	−0.30			0.66		2.64	0.96

Table 9.6 Correlation matrix for seven heavy metals in the soil at La Chaux-de-Fonds in the Swiss Jura with 364 degrees of freedom.

Log cadmium	1						
Cobalt	0.393	1					
Chromium	0.653	0.473	1				
Log copper	0.243	0.271	0.300	1			
Nickel	0.634	0.727	0.717	0.326	1		
Log lead	0.346	0.212	0.335	0.795	0.372	1	
Log zinc	0.677	0.523	0.669	0.700	0.687	0.685	1
	Cd	Co	Cr	Cu	Ni	Pb	Zn

Table 9.7 Eigenvalues of correlation matrix for seven heavy metals in the soil at La Chaux-de-Fonds.

Order	Eigenvalue	Percentage	Accumulated percentage
1	4.123	58.90	58.90
2	1.342	19.17	78.07
3	0.681	9.72	87.79
4	0.344	4.91	92.70
5	0.229	3.27	95.97
6	0.162	2.31	98.28
7	0.120	1.72	100.00

sured. Atteia *et al.* (1994) describe the sampling and analytical procedure in detail.

Table 9.5 summarizes the data from the 366 sites. It shows immediately that the frequency distributions of four of the metals—Cd, Cu, Pb and Zn—were strongly skewed, and so the data for these were transformed to their common logarithms to stabilize their variances.

Further, the correlation matrix, Table 9.6, shows some fairly strong correlations—between Co and Ni, and between Cu and Pb, for example. The general strength of correlation in the data may be judged by converting the matrix to principal components. The results are summarized in Table 9.7. The two leading principal components account for 78% of the variance in the matrix, given in the right-hand column. The correlation may then be displayed in the plane of the first two axes by computing the correlation coefficients, c_{ij}, between the principal component scores and the original variables, as follows:

$$c_{ij} = a_{ij}\sqrt{v_j/\sigma_i^2}, \qquad (9.25)$$

where a_{ij} is the ith element of the jth eigenvector, v_j is the jth eigenvalue,

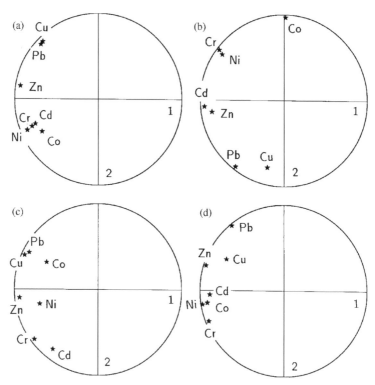

Figure 9.5 Projections of the correlations between the original (standardized) variables and the principal component scores into unit circles in the plain of the first two principal components for the heavy metals in the Swiss Jura: (a) the ordinary product-moment correlation matrix; (b) the nugget matrix; (c) the short-range matrix; (d) the long-range matrix.

and σ_i^2 is the variance of the ith original variable. We then plot these coefficients in circles of unit radius in the planes of the leading components. Figure 9.5(a) shows the result. The first axis represents the magnitude of the concentrations: large concentrations of one metal are associated with large concentrations of the others. Axis 2 spreads the metals out, and it is evident that Cu and Pb are closely associated, as are the transition metals Co, Ni and Cr. Cadmium seems to be related to these metals, while Zn lies about half-way between the two groups. The occurrence of all the points close to the circumference of the circle is one more reflection of there being only little more information in the other dimensions.

The principal component analysis has another advantage: the leading components should concentrate the information on the spatial structure. Figure 9.6 shows the variograms of the first two components. The plotted points are the experimental values and the solid lines are the fitted models. We fitted double spherical functions with nugget variances (see Chap-

Table 9.8 Coefficients of double-spherical model fitted to principal components for La Chaux-de-Fonds.

	c_0	c_1	c_2	a_1	a_2
Component 1	0.547	1.473	2.580	0.198	1.376
Component 2	0.543	0.513	0.330	0.377	1.976
Component 3	0.174	0.287	0.239	0.161	1.297

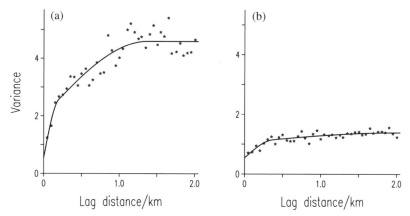

Figure 9.6 Variograms of (a) the first and (b) the second principal components of the heavy metals. The points show the experimental values and the solid lines are of the independently fitted double spherical models, the coefficients of which are listed in Table 9.10.

ter 6) to both, and the coefficients for the models are listed in Table 9.8. The nested structure of the first component is clear, with two distinct ranges, $a_1 \approx 0.2$ km and $a_2 \approx 1.3$ km.

The first principal component contains such a large proportion of the total variation that we have taken its spatial structure and its distance parameters, the two ranges, as typical of the full set of data. We set these ranges as constants, and then found the sills of the seven autovariograms and the 21 cross-variograms iteratively by the Goulard and Voltz algorithm. These sills are listed in Table 9.9. Figure 9.7 shows the autovariograms with the model fitted to them, and readers can see the full set including the cross-variograms in Webster *et al.* (1994).

The differences between the metals are striking. The spatial correlation of Cd, Pb and Cu is dominantly of short range, whereas that of Co and Ni is of long range. Somewhat surprisingly, the variogram of Cr is dominated by the short-range component. Zinc has an intermediate structure.

We can take the analysis one stage further by finding the principal components of the coregionalization matrices, as follows. The coefficients,

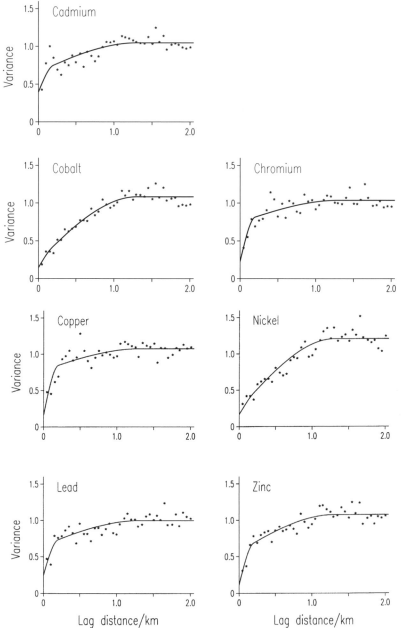

Figure 9.7   Experimental autovariograms of the seven heavy metals in the soil of the Swiss Jura shown by point symbols and the fitted model of coregionalization shown by solid lines. All of the scales have been standardized to variance equal to 1 for comparison. The coefficients are listed in Table 9.9.

Table 9.9 Coefficients, b^k, of double-spherical model of coregionalization for standardized autovariograms of the seven heavy metals in the soil at La Chaux-de-Fonds. All of the scales have been standardized to variance equal to 1 for comparison.

	Cd	Co	Cr	Cu	Ni	Pb	Zn
Nugget, b^1	0.396	0.146	0.225	0.160	0.168	0.244	0.113
Short range, b^2	0.267	0.092	0.524	0.621	0.073	0.408	0.454
Long range, b^3	0.384	0.844	0.288	0.291	0.961	0.344	0.508

Table 9.10 Nugget and structural correlation coefficients in lower triangles for seven heavy metals in the soil at La Chaux-de-Fonds.

Nugget variances						
Log cadmium	1					
Cobalt	−0.051	1				
Chromium	0.399	0.217	1			
Log Copper	0.118	−0.253	−0.137	1		
Nickel	0.347	0.197	0.446	−0.040	1	
Log lead	0.322	−0.276	0.037	0.249	0.021	1
Log zinc	0.521	−0.093	0.249	0.094	0.237	0.233	1

Short-range components						
Log cadmium	1					
Cobalt	0.167	1				
Chromium	0.358	0.050	1			
Log Copper	0.126	0.174	0.292	1		
Nickel	0.097	−0.021	0.579	0.422	1	
Log lead	0.028	0.210	0.326	0.637	0.153	1
Log zinc	0.316	0.193	0.502	0.609	0.152	0.421	1

Long-range components						
Log cadmium	1					
Cobalt	0.427	1				
Chromium	0.459	0.786	1			
Log Copper	0.237	0.578	0.292	1		
Nickel	0.578	0.766	0.499	0.302	1	
Log lead	0.363	0.412	0.127	0.327	0.481	1
Log zinc	0.485	0.719	0.398	0.370	0.837	0.461	1
	Cd	Co	Cr	Cu	Ni	Pb	Zn

b^k_{uv}, for all $u = 1, 2, \ldots, V$ and all $v = 1, 2, \ldots, V$, constitute a $V \times V$ variance–covariance matrix, $\mathbf{B}^k$, and principal components of these can be found in exactly the same way as those of any other variance–covariance

Table 9.11 Eigenvalues of structural variance–covariance matrices for La Chaux-de-Fonds.

Order	Nugget		Short range		Long range	
	Eigenvalue	Accumulated percentage	Eigenvalue	Accumulated percentage	Eigenvalue	Accumulated percentage
1	0.7498	52.20	1.5965	65.44	2.8339	78.27
2	0.4792	85.61	0.4979	85.85	0.3608	88.23
3	0.0974	92.39	0.1930	93.76	0.2546	95.26
4	0.0685	97.16	0.1018	97.93	0.1215	98.62
5	0.0406	99.99	0.0504	100.00	0.0497	99.99
6	0.0001	100.00	0.0000	100.00	0.0003	100.00
7	0.0000	100.00	0.0000	100.00	0.0000	100.00

or correlation matrix. The elements of the matrix, which are listed in Table 9.10, are converted to correlation coefficients by dividing by the square roots of the variances on the diagonal so that all variables have equal weight. The eigenvalues, v, and eigenvectors, a, are then extracted. To explore the relations among the variables we computed the correlations between the original variables and the principal components at each scale using equation (9.25), replacing σ_i^2 by the relevant b_{uu}^k.

Figure 9.5(b)–(d) shows the results for the nugget, short-range and long-range components, respectively. The first two eigenvalues account for more than 85% of the variance in all three matrices (Table 9.11), and in consequence all the points plot near the circumference. The contributions of the nugget variance appear as a scatter of points to the left of centre in Figure 9.5(b). In Figure 9.5(c), representing the short-range components, Cu and Pb are close neighbours—evidently they are closely correlated at this scale—whereas the uncorrelated Co and Ni make little contribution at this scale and so lie closer to the centre. The reverse is the case at the long range (Figure 9.5(d)), in which the strong correlation between Co and Ni is apparent.

Webster *et al.* (1994) thought that the two distinct patterns of variation might result from two distinct sources of the metals; the transition metals Co and Ni, deriving from the rocks, and Cu, Pb, Zn and perhaps Cd having been added in manure, fertilizer, sewage sludge or urban waste. They used the results to explore these possibilities.

9.6 PSEUDO-CROSS-VARIOGRAM

It will be evident from the computing formula, equation (9.10), that cross-semivariances can be calculated only from points where both variables u and v have been measured. In the examples from Broom's Barn Farm and the Jura there are few missing data, and the restriction is of little consequence. There are other situations, however, where it is difficult or even impossible to measure the two variables at the same place, as when sampling is destructive. This happens in soil monitoring. Soil material may be taken away initially for analysis and is not there subsequently, so on later occasions the soil must be measured at different places (see Papritz *et al.*, 1993, and Papritz and Webster, 1995a; 1995b). Nevertheless, one may have many observations from which to assess spatial relations and one would like to use them.

Clark *et al.* (1989) recognized the desire, and they proposed a 'pseudo-cross-variogram'. They introduced it with the following definition:

$$\gamma_{uv}^C(\mathbf{h}) = \tfrac{1}{2}E[\{Z_u(\mathbf{x}) - Z_v(\mathbf{x} + \mathbf{h})\}^2]. \tag{9.26}$$

This is unsatisfactory because, unless $\mu_u = \mu_v$, $\gamma_{uv}^C(\mathbf{h})$ is not equal to the half of the variance of the difference. Myers (1991) recognized this shortcoming and redefined the pseudo-cross-variogram as the variance:

$$\gamma_{uv}^{P}(\mathbf{h}) = \tfrac{1}{2}\operatorname{var}[Z_u(\mathbf{x}) - Z_v(\mathbf{x} + \mathbf{h})]. \tag{9.27}$$

If the means of u and v are equal then $\gamma_{uv}^{C}(\mathbf{h}) = \gamma_{uv}^{P}(\mathbf{h})$; otherwise the function defined by Clark *et al.* equals $\gamma_{uv}^{P}(\mathbf{h}) + (\mu_u - \mu_v)^2$.

For second-order stationary processes $\gamma_{uv}^{P}(\mathbf{h})$ is related to the cross-covariance function by

$$\gamma_{uv}^{P}(\mathbf{h}) = \tfrac{1}{2}\{C_{uu}(\mathbf{0}) + C_{vv}(\mathbf{0})\} - C_{uv}(\mathbf{h}). \tag{9.28}$$

Like the cross-covariance function, it is in general not symmetric in $\mathbf{h}$. It is also related to the ordinary cross-variogram by

$$\gamma_{uv}^{P}(\mathbf{h}) + \gamma_{vu}^{P}(\mathbf{h})$$
$$= \gamma_{uu}(\mathbf{h}) + \gamma_{vv}(\mathbf{h}) + 2\gamma_{uv}(\mathbf{h})$$
$$- \operatorname{cov}[\{Z_u(\mathbf{x}) - Z_v(\mathbf{x} + \mathbf{h})\}, \{Z_v(\mathbf{x}) - Z_u(\mathbf{x} + \mathbf{h})\}], \tag{9.29}$$

and for second-order stationary processes with symmetric cross-covariances

$$\gamma_{uv}^{P}(\mathbf{h}) = \gamma_{uv}(\mathbf{h}) + \tfrac{1}{2}\{C_{uu}(\mathbf{0}) + C_{vv}(\mathbf{0}) - 2C_{uv}(\mathbf{0})\}. \tag{9.30}$$

Papritz *et al.* (1993) have explored the properties of the pseudo-cross-variogram and discovered that it has rather restricted validity, though in the right conditions it can be modelled with the ordinary autovariograms and used for cokriging, and this is likely to be its main attraction. More generally, the inability to estimate the usual cross-variogram for want of comparisons between variables at lag zero is tantalizing. Papritz *et al.* (1993) suggested a way forward for situations in which the pseudo-cross-variogram is valid, but the computational load still seems prohibitive for the size of sample needed for reliable estimation.

At present we leave the reader with the pseudo-cross-variogram as a possible function to describe cross-correlation. It is far from ideal, and it seems to us preferable to plan surveys in such a way that there are always enough sites at which all the variables are or can be measured.

For further details and explanation, see Journel and Huijbregts (1978), Matheron (1979), Myers (1982), and McBratney and Webster (1983), Papritz *et al.* (1993) and Wackernagel (1994; 1995).

10
Disjunctive kriging

10.1 INTRODUCTION

Ordinary kriging is the most common form of geostatistical estimation. As described in Chapter 8, it estimates the values of regionalized variables at unsampled places, i.e. at the target points or blocks, as simple linear combinations of measured values in the neighbourhoods of those targets. The estimates are the best of their kind in the sense that they are unbiased and the variance, which is also estimated, is the minimum. Sometimes we should like to have more information than this; for instance, we may want to know, given the data, the likelihood or probability that the true values at the target points exceed some threshold. These probabilities are not linear combinations of the data. To estimate them we need more elaborate techniques that depend on the statistical distributions of the variables at the target points. The following examples illustrate where this need arises.

In developed countries, in particular, there is a desire to clean up and protect the environment. In some cases laws have been passed to limit the concentrations of certain materials in the air, water and soil. For example, the European Union has stipulated a permissible maximum for the concentration of nitrate in drinking water of 50 mg l^{-1}. This has given local authorities in England and Wales powers to prosecute farmers who cause this to be exceeded in water supplies. The Swiss federal government has specified maxima for the concentrations of heavy metals in the soil of the country (FOEFL, 1987). For cadmium and lead, as examples, they are 0.8 mg kg^{-1} and 50 mg kg^{-1}, respectively. They are guide values, but if they are exceeded then the cantonal administrations must act appropriately. The quality of the air may be judged on the amount of SO_2 it contains, and governments may again set limits to what is tolerable. If a limit, denoted z_c, is exceeded then the law-enforcement agency may order polluters to cut their emissions.

In agriculture there are similar situations. In humid temperate climates the soil tends to be acid, cropping there increases the tendency, and farmers need to apply lime to counteract it. There is often a critical value of pH that signifies the need for lime. If the soil's pH falls below that value then it is time to act by adding lime to the land. The farmer would like to know, therefore, whether the pH is less than this threshold for each point on the farm. If it exceeds the critical value then the farmer need do nothing. Farmers in drier regions often have to control salinity and alkalinity. Again there are critical values of electrical conductivity in the soil solution (for salinity) and exchangeable sodium percentage (for alkalinity), and if these are exceeded then the farmer should apply gypsum and try to leach the soluble salt out. Here the thresholds, z_c, are maxima. In other situations there are minimum recommended concentrations for certain nutrient elements in soil. This is especially true of the trace metals copper and cobalt which are essential in the diets of grazing livestock, and graziers should ensure that the herbage, and therefore the soil on which it grows, contains enough.

What is common to these situations is that true values are known only at sample points. Elsewhere the environmental protection agency, the farmer, the grazier, must estimate or predict the values, and these estimates are subject to error. Decisions, however, must be based on these estimates despite the errors. Where an estimate exceeds a threshold widely or is much less than it the decision-maker can take it at its face value and act or not as is appropriate. Difficulty arises where the estimate is close to the threshold and might result in a misjudgement that could have serious or expensive consequences, or both. For example, if the true concentration of nitrate in the ground water is less than 50 mg l^{-1}, the z_c, and the water authority estimates it as more then farmers might be constrained or fined unnecessarily, whereas if the situation is the reverse then consumers might suffer.

Similarly, if the true pH of the soil is less than 5.5, the relevant threshold for a given crop, and the farmer estimates it to be more then he will not add lime. The likely outcome is a loss of yield and profit. If on the other hand the true value exceeds 5.5 and the farmer's estimate is less then he could spend money unnecessarily on lime. If the grazier overestimates the concentration of cobalt in soil that is deficient and as a result does nothing to correct the deficiency then his sheep will not thrive and may die prematurely. If he underestimates the concentration in soil containing sufficient cobalt then he might add cobalt to the soil or to the animals' diet unnecessarily or, more expensively, have his animals' blood tested.

In such situations the land manager might attempt to remedy a soil condition that did not exist, or an agency could have a false sense of security and fail to deal with a threat that did exist. To avoid unnecessary expenditure or treatment or the risk of losing yield or perpetuating environmental damage and suffering if nothing is done, the land manager or law enforcer needs to know the risks of taking their estimates at face value.

Miners face a similar problem. At any particular time there is a price of metal and the cost of processing its ore, and there is a threshold concentration greater than which it is profitable to extract each block of rock and less than which it is not. As Journel and Huijbregts (1978) remark, '[d]ecisions are based on estimates, whereas profits depend on the true values'. Miners take financial risks when treating estimates as if they are true.

If we use linear kriging to estimate Z at the nodes of a fine grid we could examine the effect of the threshold by threading an isarithm at z_c through the grid and display the result as a map. This would show two classes: one where the estimates of Z exceed z_c and the other where they do not. As with the individual estimates, the map would be more or less in error, and there would be a risk in taking the map at its face value.

In all of these situations we need estimates of the probability, given the data, that the true values exceed (or do not exceed) the threshold, z_c, at an unsampled location $\mathbf{x}_0$. It can be expressed formally by

$$\text{Prob}[Z(\mathbf{x}_0) > z \mid z(\mathbf{x}_i); \; i = 1, 2, \ldots, N] = 1 - \text{Prob}[Z(\mathbf{x}_0) \leqslant z_c \mid z(\mathbf{x}_i)],$$

$$(10.1)$$

where N is the number of data points.

To determine the probabilities we need to know the conditional expectation or expected value at each target point, which depends on knowing the probability distribution of $Z(\mathbf{x})$. Unfortunately, the full multivariate distribution of $Z(\mathbf{x})$ is inaccessible, partly because we have only one realization and partly because the actual probability distributions depart more or less from theoretical ones.

Two solutions have been proposed to overcome this difficulty; both involve transformations of data, and both are used in practice. The simpler is indicator kriging (Journel, 1983); it needs no assumption of a theoretical distribution, and in this sense it is non-parametric. It converts a variable that has been measured on a continuous scale to several indicator variables, each taking the values 0 or 1 at the sample sites, and estimating their values elsewhere. It is appealing for these reasons. The other solution, disjunctive kriging, is due to Matheron (1976). It transforms the data to a standard normal distribution using Hermite polynomials and then compares the estimated values with the normal distribution to obtain the required probabilities.

Although indicator and disjunctive kriging are described as non-linear methods, both are linear krigings of non-linear transforms of data. Indicator kriging involves simple or ordinary kriging of indicators, and disjunctive kriging is a simple kriging of Hermite polynomials. Both lead to estimates of the probabilities that the true values exceed (or not) specified thresholds at unknown points or blocks in the neighbourhood of data. In this way they enable us to assess the risk we take by accepting the estimates at their face values.

Many case studies using the techniques have been reported. Examples of indicator kriging in mining include ones by Journel (1983) and Lemmer (1984), and in environmental protection by Bierkens and Burrough (1993a; 1993b), Journel (1988), Goovaerts (1994) and Goovaerts *et al.* (1997). Matheron developed disjunctive kriging specifically for mining, and its potential benefits for that industry are evident (Rendu, 1980; Maréchal, 1976; Rivoirard, 1994). Nevertheless, it is proving well suited for environmental protection and land management. Applications in soil science have been especially successful. Yates *et al.* (1986a; 1986b) set it in the context of soil water, and Yates and Yates (1988) used it to estimate viral contamination of soil by sewage. We have applied it to several case studies in soil science (Webster and Oliver, 1989; Wood *et al.* 1990; Webster, 1991; 1994; Oliver *et al.*, 1996).

In this chapter we describe Gaussian disjunctive kriging, but before going into detail we devote a short section to indicators in general.

10.2 THE INDICATOR APPROACH

10.2.1 Indicator coding

An indicator variable, often abbreviated to 'indicator' in geostatistical parlance, is essentially a binary variable; it is one that takes the values 1 and 0 only. Typically such variables denote presence or absence. In soil science we could score a soil sample with a 1 for earthworms if they were present and 0 if they were not. We might score the presence of roots and stones similarly.

We can also create an indicator, $\omega(\mathbf{x})$, from a continuous variable, $z(\mathbf{x})$, quite simply by scoring it 1 if $z(\mathbf{x})$ is less than or equal to a specified threshold or cut-off, z_c, and 0 otherwise:

$$\omega(\mathbf{x}) = \begin{cases} 1 & \text{if } z(\mathbf{x}) \leqslant z_c, \\ 0 & \text{otherwise.} \end{cases} \tag{10.2}$$

We thereby dissect the scale of z into two parts, one for which $z(\mathbf{x}) \leqslant z_c$ and one for which $z(\mathbf{x}) > z_c$, and assign to them the values 1 and 0, respectively. This is what is meant by disjunctive coding. If $z(\mathbf{x})$ is a realization of a random process, $Z(\mathbf{x})$, then $\omega(\mathbf{x})$ may be regarded as the realization of the indicator random function, $\Omega[Z(\mathbf{x}) \leqslant z_c]$. This is a new binary random process.

It will be convenient to abbreviate the notation somewhat for these variables to $\omega(\mathbf{x}; z_c)$ for the realization and $\Omega(\mathbf{x}; z_c)$ for the random function.

The relevance of this transformation to environmental protection is evident. If we have a threshold, z_c, for the concentration of a pollutant that may not be exceeded then the continuous random variable, $Z(\mathbf{x})$, is converted to an indicator function for which the value 1 means clean or acceptable and 0 means polluted and unacceptable. We have already mentioned examples for nitrate in drinking water in the European Union and heavy metals in the soil of Switzerland.

Converting a continuous variable to an indicator clearly loses much of the information in the original data, and it might seem prodigal to transform quantitative data in this way. There are reasons for doing it, however. In some instances the statistical distribution is such that transforming it to one that is known is difficult. This is often the case where there are many zeros in the record. There may be outliers, the effects of which we want to retain. Pollutants, for which only the outliers exceed the statutory thresholds, often fall into this class.

If this were all, however, the transformation would be of little practical significance. What makes it of value is that several thresholds can be defined and a new indicator variable created for each. Thus, if we define S thresholds as $z_{c(1)}, z_{c(2)}, \ldots, z_{c(S)}$, then we shall obtain S indicators from the data, $\omega_1, \omega_2, \ldots, \omega_S$:

$$
\begin{aligned}
\omega_1(\mathbf{x}) &= 1 \quad \text{if } z(\mathbf{x}) \leqslant z_{c(1)}, \quad \text{else } 0, \\
\omega_2(\mathbf{x}) &= 1 \quad \text{if } z(\mathbf{x}) \leqslant z_{c(2)}, \quad \text{else } 0, \\
&\vdots \\
\omega_S(\mathbf{x}) &= 1 \quad \text{if } z(\mathbf{x}) \leqslant z_{c(S)}, \quad \text{else } 0.
\end{aligned}
\tag{10.3}
$$

These may be regarded as the realizations of the corresponding random functions $\Omega_s(\mathbf{x})$, $s = 1, 2, \ldots, S$, for which

$$
\Omega_s(\mathbf{x}) = 1 \quad \text{if } Z(\mathbf{x}) \leqslant z_{c(s)}, \quad \text{else } 0.
\tag{10.4}
$$

The expectation of the indicator, $\mathrm{E}[\Omega[Z(\mathbf{x}) \leqslant z_c]]$, is the probability, $\mathrm{Prob}[z_c]$, that $Z(\mathbf{x})$ does not exceed z_c:

$$
\mathrm{Prob}[z_c] = \mathrm{Prob}[Z(\mathbf{x}) \leqslant z_c] = \mathrm{E}[\Omega[Z(\mathbf{x}) \leqslant z_c]].
\tag{10.5}
$$

This probability, $\mathrm{Prob}[Z(\mathbf{x}) \leqslant z_c]$, is the cumulative distribution

$$
\begin{aligned}
\mathrm{Prob}[Z(\mathbf{x}) \leqslant z_c] &= 1 - \mathrm{Prob}[Z(\mathbf{x}) > z_c] \\
&= G[Z(\mathbf{x}; z_c)].
\end{aligned}
\tag{10.6}
$$

Many environmental variables are multi-state characters, such as types of rock, soil and vegetation, that have more than two classes. These can also be converted to indicators by coding each class as present or absent. If we wished to distinguish podzols, brown earths, rendinas and gleys in a region then we could set up four binary variables, one for each class and code each in turn as 1 or 0. The classes are mutually exclusive, and so in this instance just one of the four would be coded 1 and the other three as 0.

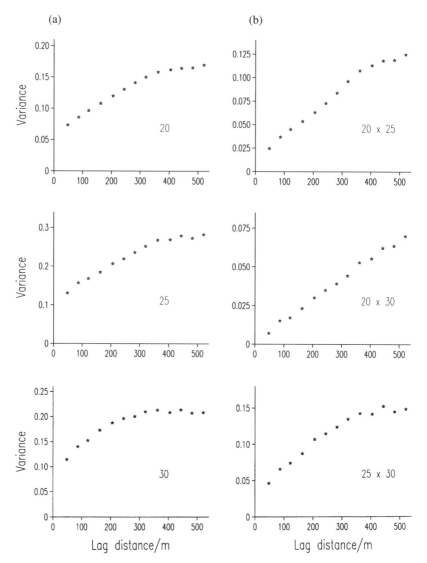

Figure 10.1 Indicator variograms of potassium at Broom's Barn Farm for thresholds of 20, 25 and 30 mg l^{-1}: (a) autovariograms; (b) cross-variograms.

10.2.2 Indicator variograms

An indicator random function has a variogram

$$\gamma^{\Omega}_{z_c}(\mathbf{h}) = \tfrac{1}{2}\mathrm{E}[\{\Omega[Z(\mathbf{x}; z_c)] - \Omega[Z(\mathbf{x} + \mathbf{h}; z_c)]\}^2], \qquad (10.7)$$

which is analogous to the variogram of a continuous variable, equation (4.12). The expected semivariances can be estimated from indicator data by

$$\hat{\gamma}^{\Omega}_{z_c}(\mathbf{h}) = \frac{1}{2m(\mathbf{h})} \sum_{i=1}^{m(\mathbf{h})} \{\omega(\mathbf{x}_i; z_c) - \omega(\mathbf{x}_i + \mathbf{h}; z_c)\}^2. \tag{10.8}$$

Figure 10.1 shows examples in the left-hand column. Further, the ordered sets, $\hat{\gamma}^{\Omega}_{z_c}(\mathbf{h})$, obtained by applying this formula with changing $\mathbf{h}$ and for all thresholds, z_c, can be modelled as described in Chapter 6.

Cross-indicator variograms

For any two thresholds, say z_u and z_v, we can define a cross-indicator variogram and estimate it by elaborating the above formulae:

$$\gamma^{\Omega}_{uv}(\mathbf{h}) = \tfrac{1}{2}\mathrm{E}[\{\Omega[Z(\mathbf{x}; z_u)] - \Omega[Z(\mathbf{x} + \mathbf{h}; z_u)]\}$$
$$\times \{\Omega[Z(\mathbf{x}; z_v)] - \Omega[Z(\mathbf{x} + \mathbf{h}; z_v)]\}]. \tag{10.9}$$

Examples of cross-variograms of indicators appear in Figures 10.1(b), 10.1(d) and 10.1(f).

Indicator covariance functions

If the processes are second-order stationary then the spatial correlations among the indicators can all be expressed in terms of covariances:

$$C^{\Omega}_{z_c}(\mathbf{h}) = \mathrm{cov}[\Omega[Z(\mathbf{x}; z_c)], \Omega[Z(\mathbf{x} + \mathbf{h}; z_c)]]$$
$$= \mathrm{E}[\Omega[Y(\mathbf{x}; z_c)]\Omega[Z(\mathbf{x} + \mathbf{h}; z_c)]] - \{\mathrm{E}[\Omega[Z(\mathbf{x}; z_c)]]\}^2. \tag{10.10}$$

Similarly, the cross-covariance at lag $\mathbf{h}$ of the indicators for thresholds z_u and z_v is

$$C^{\Omega}_{uv}(\mathbf{h}) = \mathrm{cov}[\Omega[Z(\mathbf{x}; z_u)], \Omega[Z(\mathbf{x} + \mathbf{h}; z_v)]]$$
$$= \mathrm{E}[\Omega[Z(\mathbf{x}; z_u)]\Omega[Z(\mathbf{x} + \mathbf{h}; z_v)]]$$
$$- \mathrm{E}[\Omega[Z(\mathbf{x}; z_u)]]\mathrm{E}[\Omega[Z(\mathbf{x}; z_v)]]. \tag{10.11}$$

10.3 INDICATOR KRIGING

As above, we can krige an indicator variable. So for each target point or block we compute

$$\hat{\Omega}(\mathbf{x}_0; z_c) = \sum_{i=1}^{N} \lambda_i \omega(\mathbf{x}_i; z_c), \tag{10.12}$$

where the λ_i are the weights as usual. This is the ordinary kriged estimate. The indicator is necessarily bounded, and its sample mean ($\overline{\omega}; z_c$), is usually taken as its expectation. We can therefore use simple kriging to estimate $\Omega(\mathbf{x}_0; z_c)$:

$$\hat{\Omega}(\mathbf{x}_0; z_c) = \sum_{i=1}^{N} \lambda_i \omega(\mathbf{x}_i; z_c) + \left\{ 1 - \sum_{i=1}^{N} \lambda_\omega \right\} (\overline{\omega}; z_c), \qquad (10.13)$$

with weights obtained by solving the simple kriging system

$$\sum_{i=1}^{N} \lambda_i \gamma^\Omega(\mathbf{x}_i, \mathbf{x}_j; z_c) = \gamma^\Omega(\mathbf{x}_0, \mathbf{x}_j; z_c) \qquad \text{for } j = 1, 2, \ldots, N, \qquad (10.14)$$

where $\gamma^\Omega(\mathbf{x}_i, \mathbf{x}_j; z_c)$ is the indicator semivariance between the ith and jth sampling points at threshold z_c and $\gamma^\Omega(\mathbf{x}_0, \mathbf{x}_j; z_c)$ is the semivariance of the indicator between the target point $\mathbf{x}_0$ and point $\mathbf{x}_j$ for the same threshold. As when kriging continuous variables, we can replace N by $n \ll N$ in the neighbourhood of $\mathbf{x}_0$.

The result is a value lying between 0 and 1 (with exceptions because the kriging minimizes the variance without any constraint on the estimates it returns). Such a value is effectively the probability, given the data, that the true value is 1, i.e.

$$\text{Prob}[\Omega(\mathbf{x}_0; z_c) = 1 \mid \omega(\mathbf{x}_i), \; i = 1, 2, \ldots, n] = F\{\mathbf{x}_0 \mid (n)\}, \qquad (10.15)$$

where we use (n) to mean all the data in the particular neighbourhood. The quantity $F\{\mathbf{x}_0 \mid (n)\}$ denotes the conditional or 'posterior' probability that $\Omega(\mathbf{x}_0)$ is 1.

If now we return to our problem, namely to estimate the probability, given data, that the true value of Z at an unsampled place $\mathbf{x}_0$ does not exceed z_c, then we can write

$$\text{Prob}[Z(\mathbf{x}_0) \leqslant z_c \mid z(\mathbf{x}_i); \; i = 1, 2, \ldots, N]$$
$$= 1 - \text{Prob}[Z(\mathbf{x}_0) > z_c \mid z(\mathbf{x}_i); \; i = 1, 2, \ldots, N]. \qquad (10.16)$$

Notice that the two sides of equation (10.16) are complementary.

At first sight it might seem that the way to tackle the problem is to transform the data to indicators determined by the particular threshold. However, we soon see that an individual probability estimated in this way is crude. Much of the rich information in the original data has been lost by dissecting the scale into just two classes.

This loss can be made good to a large extent by repeating the process for several thresholds in the range of Z and constructing a cumulative distribution function, conditional on the data, for each target point by accumulating the $\hat{F}(\mathbf{x}_0; z_s)$, $s = 1, 2, \ldots, S$.

The procedure is somewhat tedious because the variograms for all the thresholds must be computed and modelled. Furthermore, because the

$\hat{F}(\mathbf{x}_0; z_s)$ for the different z_s are computed independently of one another, there is no guarantee that they will sum to 1, or that the cumulative function will increase monotonically, or that the estimated probabilities will lie in the range 0 to 1. Some adjustment of the results may therefore be needed to ensure that the bounds are honoured and the order relations maintained. Nevertheless, an empirical distribution function can be obtained and then used to refine the estimate of the conditional probability that $Z(\mathbf{x}_0) \leqslant z_c$.

Goovaerts (1997) describes the procedure fully and illustrates it with examples using the computer programs in GSLIB (Deutsch and Journel, 1992), while Olea (1999) devotes a section of his book to the topic. We shall not repeat the detail here.

10.4 DISJUNCTIVE KRIGING

Disjunctive kriging provides another way of estimating an indicator transform of continuous data. It does so without losing information, though requiring rather stronger assumptions than does indicator kriging as described above. It may take several forms (see Rivoirard, 1994), the most common of which is Gaussian disjunctive kriging and the one we describe.

10.4.1 Assumptions of Gaussian disjunctive kriging

The assumptions underlying Gaussian disjunctive kriging are as follows. First, $z(\mathbf{x})$ is a realization of a second-order stationary process $Z(\mathbf{x})$ with mean μ, variance σ^2 and covariance function $C(\mathbf{h})$. The underlying variogram must therefore be bounded. Second, the bivariate distribution for the $n + 1$ variates, i.e. for each target site and the sample locations in its neighbourhood, is known and is stable throughout the region. If the distribution of $Z(\mathbf{x})$ is normal (Gaussian) and the process is second-order stationary then we can assume that the bivariate distribution for each pair of locations is also normal. Each pair of variates has the same bivariate density, and this density function is determined from the spatial autocorrelation coefficient. These assumptions allow the conditional expectations to be written in terms of the autocorrelation coefficients, as we shall show.

The variable $Z(\mathbf{x})$ is spatially continuous, so that in going from a small value at one place to a large one elsewhere it must pass through intermediate values *en route*. It is an example of a *Gaussian diffusion process*. One test of this assumption is to compare the variograms of the indicators for several thresholds within the bounds of the measured z. The cross-indicator variograms should be more 'structured' than the autovariograms. Figure 10.1 shows this to be so for potassium at Broom's Barn Farm.

10.4.2 Hermite polynomials

The requirement of normality is a strong one that is rarely met in practice, even though many environmental properties seem approximately normal. The first task therefore is to transform an actual distribution of $Z(\mathbf{x})$, which may have almost any form, to a standard normal one, $Y(\mathbf{x})$, such that

$$Z(\mathbf{x}) = \Phi[Y(\mathbf{x})]. \tag{10.17}$$

This can be done using Hermite polynomials.

We recall, from Chapter 2, the equation of the standard normal distribution with probability density given by

$$g(y) = \frac{1}{\sqrt{2\pi}} \exp\left(-\frac{y^2}{2}\right). \tag{10.18}$$

Hermite polynomials are related to this function and are defined by Rodrigues's formula as

$$H_k(y) = \frac{1}{\sqrt{k!}\,g(y)} \frac{\mathrm{d}^k g(y)}{\mathrm{d}y^k}. \tag{10.19}$$

Here k is the degree of the polynomial, taking values $0, 1, \ldots$, and $1/\sqrt{k!}$ is a standardizing factor (Matheron, 1976). The first two Hermite polynomials, i.e. for $k = 0$ and $k = 1$, are

$$H_0(y) = 1, \tag{10.20}$$
$$H_1(y) = -y; \tag{10.21}$$

and thereafter the higher-order polynomials obey the recurrence relation

$$H_k(y) = -\frac{1}{\sqrt{k}} y H_{k-1}(y) - \sqrt{\frac{k-1}{k}} H_{k-2}(y). \tag{10.22}$$

So the polynomials can be calculated up to any order for a standard normal distribution.

The Hermite polynomials are orthogonal with respect to the weighting function $\exp(-y^2/2)$ on the interval $-\infty$ to $+\infty$. They are independent components of the normal distribution of ever increasing detail.

Almost any function of $Y(\mathbf{x})$ can be represented as the sum of Hermite polynomials:

$$f\{Y(\mathbf{x})\} = f_0 H_0\{Y(\mathbf{x})\} + f_1 H_1\{Y(\mathbf{x})\} + f_2 H_2\{Y(\mathbf{x})\} + \cdots, \tag{10.23}$$

and since the Hermite polynomials are orthogonal

$$E[f\{Y(\mathbf{x})\}H_k\{Y(\mathbf{x})\}] = E\left[H_k\{Y(\mathbf{x})\}\sum_{l=0}^{\infty}f_lH_l\{Y(\mathbf{x})\}\right]$$

$$= \sum_{l=0}^{\infty}f_lE[H_l\{Y(\mathbf{x})\}H_k\{Y(\mathbf{x})\}]$$

$$= f_k. \tag{10.24}$$

This enables us to calculate the coefficients ϕ_k of $\Phi[Y(\mathbf{x})]$ in equation (10.17) as

$$Z(\mathbf{x}) = \Phi[Y(\mathbf{x})]$$

$$= \phi_0H_0\{Y(\mathbf{x})\} + \phi_1H_1\{Y(\mathbf{x})\}\phi_2H_2\{Y(\mathbf{x})\} + \cdots$$

$$= \sum_{k=0}^{\infty}\phi_kH_k\{Y(\mathbf{x})\}. \tag{10.25}$$

The transform is also invertible, which means that the results can be expressed in the same units as the original measurements.

Determining the Hermite coefficients

To determine the coefficients of the transformation, ϕ_k, for a particular set of data we proceed as follows. We arrange the N data in ascending order:

$$z_1 < z_2 < z_3 < \cdots < z_N,$$

and we denote their relative frequencies by

$$q_1, q_2, q_3, \ldots, q_N,$$

such that the sum of the frequencies is 1: $\sum_{i=1}^{N}q_i = 1$. Their cumulative frequencies are

$$F(z_1) = P[Z(\mathbf{x}) < z_1] = 0,$$

$$F(z_2) = P[Z(\mathbf{x}) < z_2] = q_1,$$

$$F(z_3) = P[Z(\mathbf{x}) < z_3] = q_1 + q_2,$$

$$\vdots$$

$$F(z_i) = P[Z(\mathbf{x}) < z_i] = \sum_{j=1}^{i-1}q_j,$$

$$\vdots$$

$$F(z_N) = P[Z(\mathbf{x}) < z_N] = 1 - q_N.$$

The cumulative frequencies have equivalents on the standard normal distribution:

$$F(z_i) = G(y_i). \tag{10.26}$$

Thus

$$F(z_{i+1}) - F(z_i) = G(y_{i+1}) - G(y_i), \qquad (10.27)$$

and so

$$\text{Prob}[z_i \leqslant Z(\mathbf{x}) < z_{i+1}] = \text{Prob}[y_i \leqslant Y(\mathbf{x}) < y_{i+1}],$$
$$\text{Prob}[Z(\mathbf{x}) = z_i] = \text{Prob}[y_i \leqslant Y(\mathbf{x}) < y_{i+1}]. \qquad (10.28)$$

In words, $Z(\mathbf{x})$ equals z_i when the standard normal equivalent lies between y_i and y_{i+1}. We can then determine the transformation coefficients as follows:

$$\phi_0 = E[\Phi\{Y(\mathbf{x})\}] = E[Z(\mathbf{x})] = \sum_{i=1}^{N} q_i z_i, \qquad (10.29)$$

and thereafter

$$\phi_k = E[Z(\mathbf{x})H_k\{Y(\mathbf{x})\}]$$
$$= \int_{-\infty}^{+\infty} \Phi(y)H_k(y)g(y)\,\mathrm{d}y$$
$$= \sum_{i=1}^{N} \int_{y_i}^{y_{i+1}} z_i H_k(y)g(y)\,\mathrm{d}y$$
$$= \sum_{i=1}^{N} z_i \left[\frac{1}{\sqrt{k}} H_{k-1}(y_{i+1})g(y_{i+1}) - \frac{1}{\sqrt{k}} H_{k-1}(y_i)g(y_i) \right]$$
$$= \sum_{i=2}^{N} (z_{i-1} - z_i) \frac{1}{\sqrt{k}} H_{k-1}(y_i)g(y_i) \qquad (10.30)$$

because $g(y_0) = g(-\infty) = 0$, and $g(y_{N+1}) = g(+\infty) = 0$ also.

10.4.3 Disjunctive kriging for a Hermite polynomial

Since the polynomials are orthogonal any pair of values, $Y(\mathbf{x})$ and $Y(\mathbf{x}+\mathbf{h})$, drawn from a bivariate normal distribution with correlation coefficient ρ has expectation

$$E[H_k\{Y(\mathbf{x})\} \mid Y(\mathbf{x} + \mathbf{h})\}] = \rho^k(\mathbf{h})H_k\{Y(\mathbf{x} + \mathbf{h})\}. \qquad (10.31)$$

The covariance between two functions of Y at $\mathbf{x}$ and $\mathbf{x} + \mathbf{h}$ is

$$\text{cov}[H_k\{Y(\mathbf{x})\}, H_l\{Y(\mathbf{x} + \mathbf{h})\}]$$
$$= E[H_k\{Y(\mathbf{x})\}H_l\{Y(\mathbf{x} + \mathbf{h})\}]$$
$$= E[H_k\{Y(\mathbf{x})\}E[H_l\{Y(\mathbf{x} + \mathbf{h})\}] \mid Y(\mathbf{x} + \mathbf{h})]$$
$$= \rho^k(\mathbf{h})E[H_k\{Y(\mathbf{x})\}H_l\{Y(\mathbf{x})\}]. \qquad (10.32)$$

When $k = l$ this equation gives the covariance of $H_k\{Y(\mathbf{x})\}$, which equals $\rho^k(\mathbf{h})$, since $Y(\mathbf{x})$ is a standard normal variate. The correlation coefficient, $\rho(\mathbf{h}), \mathbf{h} \neq \mathbf{0}$, must lie between -1 and $+1$; so $\rho^k(\mathbf{h})$ rapidly approaches 0 as k increases, and the spatial dependence in $H_k\{Y(\mathbf{x})\}$ declines to nothing, i.e. $H_k\{Y(\mathbf{x})\}$ becomes pure nugget.

Any pair of Hermite polynomials is spatially independent, so they are the independent factors of the bivariate normal model. By kriging them separately the estimates have only to be summed to give the disjunctive kriging estimator:

$$\hat{Z}^{\mathrm{DK}}(\mathbf{x}) = \phi_0 + \phi_1 \hat{H}_1^{\mathrm{K}}\{Y(\mathbf{x})\} + \phi_2 \hat{H}_2^{\mathrm{K}}\{Y(\mathbf{x})\} + \cdots. \tag{10.33}$$

So, if we have n points in the neighbourhood of $\mathbf{x}_0$ where we want an estimate, we estimate the Hermite polynomials by

$$\hat{H}_k^{\mathrm{K}}\{Y(\mathbf{x}_0)\} = \sum_{i=1}^{n} \lambda_{ik} H_k\{Y(\mathbf{x}_i)\}, \tag{10.34}$$

and we insert them into equation (10.33). The λ_{ik} are the kriging weights, which are found by solving the equations for simple kriging because we can assume the mean is known:

$$\sum_{i=1}^{n} \lambda_{ik} \operatorname{cov}[H_k\{Y(\mathbf{x}_j)\}, H_k\{Y(\mathbf{x}_i)\}]$$
$$= \operatorname{cov}[H_k\{Y(\mathbf{x}_j)\}, H_k\{Y(\mathbf{x}_0)\}] \quad \text{for all } j, \tag{10.35}$$

or alternatively,

$$\sum_{i=1}^{n} \lambda_{ik} \rho^k(\mathbf{x}_i - \mathbf{x}_j) = \rho^k(\mathbf{x}_j - \mathbf{x}_0) \quad \text{for all } j, \tag{10.36}$$

from equation (10.31). In particular, the procedure enables us to estimate $Z(\mathbf{x}_0)$ by

$$\hat{Z}(\mathbf{x}_0) = \Phi\{\hat{Y}(\mathbf{x}_0)\} = \phi_0 + \phi_1[\hat{H}_1^{\mathrm{K}}\{y(\mathbf{x}_0)\}] + \phi_2[\hat{H}_2^{\mathrm{K}}\{y(\mathbf{x}_0)\}] + \cdots. \tag{10.37}$$

10.4.4 Estimation variance

The kriging variance of $\hat{H}_k\{Y(\mathbf{x})\}$ is

$$\sigma_k^2(\mathbf{x}_0) = 1 - \sum_{i=1}^{n} \lambda_{ik} \rho^k(\mathbf{x}_i - \mathbf{x}_0), \tag{10.38}$$

and the disjunctive kriging variance of $\hat{f}[Y(\mathbf{x}_0)]$ is

$$\sigma_{\mathrm{DK}}^2(\mathbf{x}_0) = \sum_{k=1}^{\infty} f_k^2 \sigma_k^2(\mathbf{x}_0). \tag{10.39}$$

10.4.5 Conditional probability

Once the Hermite polynomials have been estimated at a target point we can estimate the conditional probability that the true value there exceeds the critical value, z_c. The transformation $Z(\mathbf{x}) = \Phi[Y(\mathbf{x})]$ means that z_c has an equivalent y_c on the standard normal scale. Since the two scales are monotonically related their indicators are the same:

$$\Omega[Z(\mathbf{x}) \leqslant z_c] = \Omega[Y(\mathbf{x}) \leqslant y_c]. \qquad (10.40)$$

For $\Omega[Y(\mathbf{x}) > y_c]$, which is the complement of $\Omega[Y(\mathbf{x}) \leqslant y_c]$, the kth Hermite coefficient is

$$\begin{aligned}
f_k &= \int_{-\infty}^{+\infty} \Omega[y \leqslant y_c] H_k(y) g(y) \, \mathrm{d}y \\
&= \int_{-\infty}^{y_c} H_k(y) g(y) \, \mathrm{d}y.
\end{aligned} \qquad (10.41)$$

The coefficient for $k = 0$ is the cumulative distribution to y_c,

$$f_0 = G(y_c),$$

and for larger k,

$$f_k = \frac{1}{\sqrt{k}} H_{k-1}(y_c) g(y_c). \qquad (10.42)$$

The indicator can be expressed in terms of the cumulative distribution and the Hermite polynomials:

$$\Omega[Y(\mathbf{x}) \leqslant y_c] = G(y_c) + \sum_{k=1}^{\infty} \frac{1}{\sqrt{k}} H_{k-1}(y_c) g(y_c) H_k\{Y(\mathbf{x})\}. \qquad (10.43)$$

Its disjunctive kriging estimate is obtained by

$$\hat{\Omega}^{\mathrm{DK}}[y(\mathbf{x}_0) \leqslant y_c] = G(y_c) + \sum_{k=1}^{L} \frac{1}{\sqrt{k}} H_{k-1}(y_c) g(y_c) \hat{H}_k^{\mathrm{K}}\{y(\mathbf{x}_0)\}, \qquad (10.44)$$

where L is some small number. The kriged estimates $\hat{H}_k^{\mathrm{K}}\{y(\mathbf{x}_0)\}$ approach 0 rapidly with increasing k, and so summation need extend over only few terms even though the $(1/\sqrt{k}) H_{k-1}(y_c) g(y_c)$ are considerable. Of course, this is the same as $\hat{\Omega}^{\mathrm{DK}}[z(\mathbf{x}_0) \leqslant z_c]$. Conversely, to obtain the probability of excess we can compute

$$\begin{aligned}
\hat{\Omega}^{\mathrm{DK}}[z(\mathbf{x}_0) > z_c] &= \hat{\Omega}^{\mathrm{DK}}[y(\mathbf{x}_0) > y_c] \\
&= 1 - G(y_c) - \sum_{k=1}^{L} \frac{1}{\sqrt{k}} H_{k-1}(y_c) g(y_c) \hat{H}_k^{\mathrm{K}}\{y(\mathbf{x}_0)\}.
\end{aligned}$$
$$(10.45)$$

10.4.6   Change of support

In describing disjunctive kriging above we have treated each target as a 'point' with the same support as the data. The simple kriging equations are readily modified to estimate the Hermite polynomials and hence $Z(\mathbf{x})$ over larger blocks B by replacing the covariances on their right-hand sides with block averages. The result is a block kriging, i.e. an estimate of the average value of Z within a target block, say $Z(B)$. It will also produce an estimate of the average probability that $Z(\mathbf{x}) \leqslant z_c$ in B, but note that this probability is not same as the probability that the average of Z in B is less than or equal to z_c.

As we saw above, in Chapter 4, the distribution of a spatially correlated variable changes as the support changes. In particular, the variance diminishes as the support increases and this is evident in the regularized variogram of Figure 4.7. If we are to estimate the conditional probabilities that block averages exceed z_c then we need to take into account the larger support and to model the change of support. Webster (1991) summarizes the theory and illustrates it with an example from agricultural science, and Rivoirard (1994) treats it more didactically, again with an illustration using the same data. The subject is beyond the scope of this book, but you can read about the theory and technique in the two works cited.

10.5   CASE STUDY

To illustrate the method and to enable the results of disjunctive kriging to be compared with those of ordinary kriging in Chapter 8, we use the data on exchangeable potassium from the soil survey of Broom's Barn Farm. Chapter 2 describes them in full, and here we repeat only the most salient features. Table 10.1 summarizes the statistics of the data and the transforms to standard normal deviates using Hermite polynomials. It includes the summary for the common logarithms for comparison.

Figure 10.2 shows the general nature of the problem. Figure 10.2(a), is the cumulative distribution of exchangeable K as observed. For a defined threshold concentration, z_c, we should like to know its equivalent on the standard normal curve, because then we can calculate confidence limits. We suppose for illustration that it is 20 mg l^{-1} of soil. From Figure 10.2(a) we see that the cumulative sum is approximately 0.24. Tracing this value across by the horizontal dashed lines to the standard normal distribution on Figure 10.2(b), we see that its equivalent normal deviate is -0.69, shown by the vertical dashed line there. The first task therefore is to transform the data to a standard normal distribution so that we have the equivalences for all reasonable values of z.

Table 10.1   Summary statistics.

Statistic	K	$\log_{10} K$	Hermite-transformed K
Mean	26.3	1.40	0.074 0
Median	25.0	1.40	0.104
Standard deviation	9.04	0.134	0.974
Variance	81.706	0.018 00	0.949 5
Skewness	2.03	0.39	−0.03
Kurtosis	9.45	0.57	0.07
Deficiency threshold	25.0	1.40	0.104

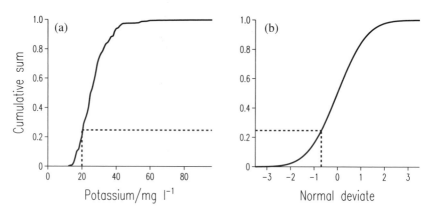

Figure 10.2   The cumulative distribution (a) of potassium and (b) of a standard normal distribution. The vertical dashed line in (a) is for a threshold of 20 mg l^{-1}, and the others show how it equates in (b); see text.

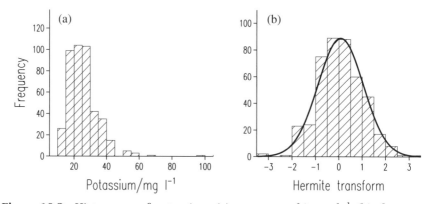

Figure 10.3   Histograms of potassium: (a) as measured in mg l^{-1}; (b) after transformation by Hermite polynomials with the curve of the normal distribution fitted.

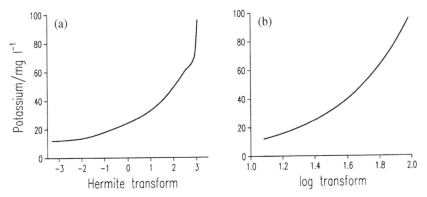

Figure 10.4 Transform functions of potassium at Broom's Barn Farm: (a) for Hermite polynomials; (b) for logarithms.

The distribution on the original scale (mg l^{-1}) is strongly skewed, $\hat{\beta}_1 = 2.04$ (Figure 10.3(a) and Table 10.1). Taking logarithms removes most of the skewness, with $\hat{\beta}_1 = 0.39$, as shown in Figure 2.1(b).

Transforming using Hermite polynomials up to order 7 is more effective, giving approximate standard normal deviates. The mean and variance depart somewhat from 0 and 1, respectively. The skewness is virtually nil (−0.03), as is the kurtosis (0.07) (see Figure 10.3(b) and Table 10.1). Figure 10.4(a) shows the transform function with the measured values plotted against the Hermite transformed ones. The graph is concave upwards, resulting from the positive skewness of the data. For a normal distribution the transform function would be a straight line; the departure from this is a measure of the non-normality. We show the logarithmic transformation function in Figure 10.4(b) for comparison.

Figure 10.5 shows other features of the transformation, again with those of the logarithms alongside for comparison. In Figure 10.5(a)–(b) are the cumulative distributions with $G(y)$ plotted against y, the transformed values. Both are characteristically sigmoid, as expected for data from a normal distribution. In Figure 10.5(c)–(d) we have plotted the normal equivalent deviates of the cumulative distributions against y. The normal equivalent deviate is the area beneath a curve of the standard normal pdf from $-\infty$ to $g(y)$, equivalent to $G(y)$. For a normal distribution this function plots as a straight line. For the Hermite transformation of potassium it is straight, apart from local fluctuation. For the logarithms, however, there is still detectable curvature.

We mentioned above our assumption that $z(\mathbf{x})$ is the outcome of a Gaussian diffusion process for which the cross-variograms of the indicators should be more structured than the autovariograms. To check that the exchangeable K conforms we computed the relevant variograms for $K > K_c$ for $c = 20, 25$ and 30 mg l^{-1}, which correspond closely to the quar-

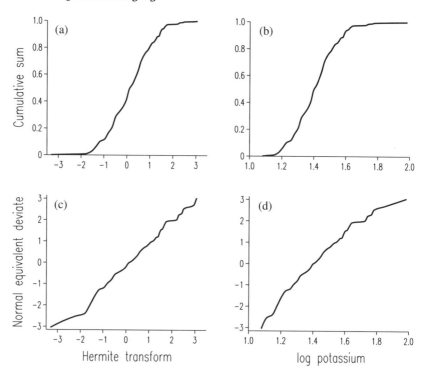

Figure 10.5 Cumulative distributions of potassium at Broom's Barn Farm: (a) the cumulative sum for the Hermite transform, and (b) for the logarithms; (c) and (d) the cumulants plotted as normal equivalent deviates.

tiles of the cumulative distribution; they are the cumulants 0.24, 0.51 and 0.75, respectively. The results are shown in Figure 10.1 with the autoindicator variograms on the left and the cross-variograms on the right. We have not fitted models to them, but quite evidently the latter are more structured; any curve fitted closely to the experimental values will project on to the ordinate near the origin, whereas all three autovariograms will have substantial nugget variances.

We computed the experimental variogram from the Hermite-transformed values and fitted an isotropic spherical model to it:

$$\hat{y}(h) = 0.216 + 0.784\,\mathrm{sph}(434),   (10.46)$$

in which sph(434) indicates the spherical function with a range of 434 m. Figure 10.6 shows the experimental values as points and the fitted model as a solid line. Using this model and the transformed values, we estimated the concentrations of potassium at the nodes of a square grid at 10 m intervals by punctual kriging of the Hermite polynomials, as described above.

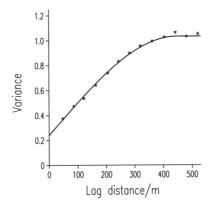

Figure 10.6   Variogram of potassium after transformation by Hermite polynomials.

For cereal crops at the time the survey was made, a critical value for readily exchangeable potassium was 25 mg K l^{-1} soil; this was the threshold below which the Ministry of Agriculture, Fisheries and Food (1986) recommended farmers to fertilize cereal crops. We computed the conditional probabilities of the values being less than or equal to this threshold at the same grid nodes.

Figure 10.7 is the map of the disjunctively kriged estimates of exchangeable K. As it happens in this instance, it is little different from the map made by lognormal kriging (Figure 8.22), because the transform functions are similar. We can see this by plotting the disjunctively and lognormally kriged estimates against each other, as in Figure 10.8(a). There is little scatter in the points from the solid line of perfect correlation on the graph, and the correlation is $r = 0.994$. Figure 10.8(b) is the scatter diagram of the disjunctively kriged estimates plotted against the kriging variance. This shows clearly the effect of the nugget variance in punctual kriging. The nugget variance sets a lower limited to the precision of the estimates, and this is evident in the horizontal line at a kriging variance of about 25 $(mg\, l^{-1})^2$.

In addition disjunctive kriging enables us to map the estimated conditional probabilities of deficiency or excess from the same set of target points. Figures 10.9(a) and 10.9(b) are maps of the probabilities for thresholds of 25 mg l^{-1} and 20 mg l^{-1}, respectively.

In Figure 10.8(c) we have plotted the conditional probabilities that the exchangeable K $\leqslant$ 25 mg l^{-1} against the disjunctively kriged estimates. It is evident from this graph that some of the estimates exceeding the threshold have associated with them fairly large probabilities of deficiency; evidently we should not judge the likelihood of deficiency from the estimates alone. You may also notice that some of the points on the graph lie out-

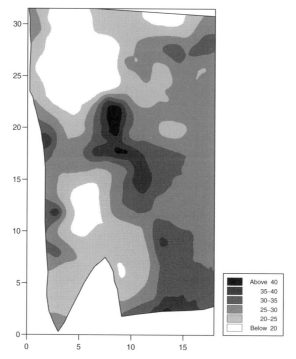

Figure 10.7   Map of exchangeable potassium at Broom's Barn Farm, estimated by disjunctive kriging.

side the bounds of 0 and 1 for the probabilities. This is because they are themselves estimates.

In environmental management we are often concerned with the probabilities that some substance exceeds a threshold. If potassium were a pollutant then we might plot the probabilities of its exceeding the threshold of 25 mg l^{-1}. We should then obtain Figure 10.8(d), the inverse of Figure 10.8(c).

In a situation concerning deficiency the farmer would fertilize where the map showed exchangeable K to be less than 25 mg l^{-1}, the pale grey and white areas of Figure 10.7. However, the farmer would not want to risk losing yield where the estimated concentration of K is more than the threshold and the probability of deficiency is moderate. If he were prepared to set the maximum risk at a probability of 0.3 then he should fertilize the areas in Figure 10.9(a) where the probability is greater, i.e. areas of medium and dark grey and black. The area is considerable. When the map of probabilities is compared with that of the estimates it is clear that the farmer could risk loss of yield by taking the estimates at face value—the area requiring fertilizer is considerably greater than that where K ⩽ 25 mg l^{-1}.

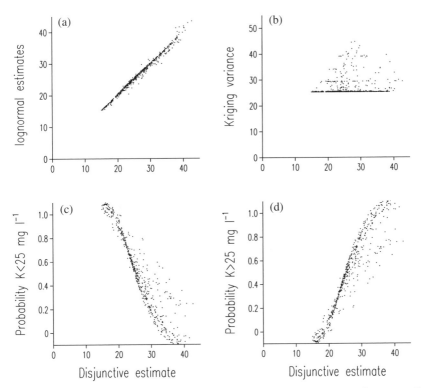

Figure 10.8 Scatter of: (a) disjunctively kriged estimates against lognormally kriged ones; (b) estimates obtained by disjunctive kriging against their estimation variances; (c) estimated probabilities of deficiency ($\leqslant 25$ mg l^{-1}) against estimates; (d) estimated probabilities of excess (> 25 mg l^{-1}) against estimates.

10.6 OTHER CASE STUDIES

Wood *et al.* (1990) described an application of disjunctive kriging to estimating the salinity of the soil in the Bet Shean Valley to the west of the River Jordan in Israel. The combination of climate, irrigation and smectite clay soil has resulted in significant concentrations of sodium salts in the topsoil. In general, salinity limits the range of crops that can be grown as well as reducing the yields of those that can tolerate it. A critical threshold, z_c, of electrical conductivity (EC) is 4 mS cm^{-1}: it is widely recognized as marking the onset of salinization of the soil. The principal crops that are affected by too much salt in the valley are lucerne, wheat and dates. The losses of yield of lucerne and dates become serious when this threshold is exceeded. Winter wheat, however, will still grow, but when the threshold is exceeded it germinates poorly.

The EC of the soil solution in November before the onset of winter rain is the most telling, and it was measured at some 200 points in a part of

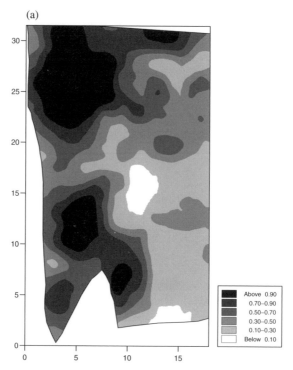

(a)

Figure 10.9   Maps of probabilities of potassium deficiency at Broom's Barn Farm at (a) a threshold of 25 mg l^{-1}.

the valley at that time to indicate salinity. The EC was then estimated at the nodes of a fine grid and mapped. The conditional probabilities of the ECs exceeding 4 mS cm^{-1} were also determined and mapped, and in the event they exceeded 0.3 over most of the region. There was a moderate risk of salinity in most of the region. Farmers would find it too costly to remediate the entire area, but they could use the map of probabilities as a guide for deciding on the priority of areas for remediation.

Webster and Oliver (1989), Webster and Rivoirard (1991) and Webster (1994) used the data from the original survey by McBratney *et al.* (1982) of copper and cobalt in the soil of south-east Scotland to study the merit and relevance of disjunctive kriging in agriculture. Deficiencies of copper and cobalt in the soil of the region cause poor health in grazing sheep and cattle there. The critical value, z_c, for copper in the soil is 1 mg kg^{-1} and for cobalt 0.25 mg kg^{-1}. Data from some 3500 sampling points were available, and from them they computed the probabilities of the soil's being deficient in the two trace metals by disjunctive kriging. The concentration of copper exceeded the 1 mg kg^{-1} threshold almost everywhere. The concentration is near to the threshold in only small parts of the region where

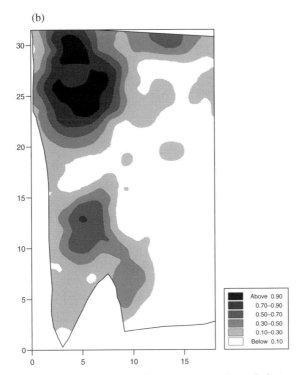

(b)

Figure 10.9   (*Cont.*) Maps of probabilities of potassium deficiency at Broom's Barn Farm at (b) a threshold of 20 mg l⁻¹.

the estimated probability of deficiency was typically in the range 0.2–0.3. For cobalt, however, for which the mean concentration was almost exactly equal to the threshold of 0.25 mg kg⁻¹, the estimates for approximately half the region were less than the threshold with an estimated probability of deficiency greater than 0.5, and elsewhere most of the computed probabilities exceeded 0.2. The potential loss of thrift in the animals and therefore profit to the farmer is considerable, whereas preventive measures such as supplementary cobalt in the animals' feed or additions in the fertilizer are cheap. In these circumstances the farmer would be advised to take one of these courses of action where the probability of deficiency exceeded 0.2.

Maps of probabilities also help environmental scientists to design programmes of remediation for areas considered to be polluted. Once the users have decided what risks they are prepared to take, the scientist can use such maps to recommend suitable action. If there are strictly limited funds for remediation the map of probabilities enables them to assign priorities for action; the parts of the region where the probabilities are greatest can be tackled first.

Von Steiger *et al.* (1996) estimated the concentrations of heavy metals in polluted soil in part of north-east Switzerland by disjunctive kriging. The soil contained lead in excess of the Swiss Federal Guide value of 50 mg kg^{-1}. The probabilities exceeded 0.3 to the north and east of the town of Weinfelden, suggesting that these areas should be monitored to ensure that the burden in the soil does not exceed the existing concentrations.

10.7 SUMMARY

The principles described in this chapter can be applied to various substances in the environment, whether they are nutrients that might be deficient or heavy metals and xenobiotics, excesses of which are toxic. The probabilities of exceeding specific thresholds enable the risk of inaction to be assessed quantitatively. Disjunctive kriging, in particular, provides environmental analysts with a useful decision-making tool, especially where failure to act could result in litigation, damage to health or loss of revenue. Assessing this risk is now feasible in an optimal way.

Appendix A
Aide-mémoire for
spatial analysis

A.1 INTRODUCTION

This appendix summarizes the steps that a scientist should take in a geostatistical analysis of survey data, beginning with error detection, summary statistics, exploratory data analysis, the variogram and its modelling, kriging, and mapping. In many instances data from remote imaging require the same treatment, and where that is so we mention it.

A.2 NOTATION

The notation is the same as used in the main text, but we repeat it here for completeness. The geographic coordinates of the sampling points and target points for prediction are denoted x_1 for eastings (or across the map from left to right) and x_2 for northings (or from bottom to top on the map). The pair $\{x_1, x_2\}$ are given the symbol $\mathbf{x}$ in vector notation. The variates are denoted $z_1, z_2, \ldots$, and the measured values are denoted $z(\mathbf{x}_i)$ for $i = 1, 2, \ldots$, for any one variate.

A.3 SCREENING

Few large files of data are free of mistakes caused by instrumental malfunction and human error. When you receive data, whether from the field or laboratory or from remote scanners, check for such mistakes.

Position. Examine the positions of the data in relation to the bounds of the region. Plot them on a map, known as a 'posting'.

- Do all the points lie within the region?
- Are there sampling points in the sea when they should be on the land? If so why?
- Have the coordinates been reversed inadvertently so that northings precede eastings, i.e. the x_2 precede the x_1?

- Do the points fill approximately the region?
- Are the coordinates properly scaled?

Measurements.   Screen the measured values, $z(\mathbf{x}_1), z(\mathbf{x}_2), \ldots$. Pass them through a program that compares each value in the file against the minimum and maximum possible of the scale and flags any that lie outside these bounds. Print out the minimum and maximum for each z and check that they are sensible.

A.4   HISTOGRAM AND SUMMARY

For each variate compute a histogram and plot it, ensuring that all classes are of the same width. Examine it for outliers, i.e. individuals or small groups that are isolated from the main body of data. In addition, if you prefer box-plots, compute them for the zs, to show outliers.

Outliers.   If there are outliers identify their positions on the map.

- Are they mistakes? If not what do they represent?
- Are they part of the 'target population?

If not (e.g. water when you are interested only in land) then replace the recorded values by a symbol to indicate missing or not applicable.

The statistical treatment of outliers is a complex subject, and if you wish to retain outliers in your analysis then consult a statistician.

Frequency distribution.   Study the shape of the histogram.

- Has it more than one peak?

If so, the scene or region almost certainly contains at least two distinct populations, e.g. land and water, farmland and forest.

- Is the distribution symmetric?
- If it is skewed is the longer tail towards the small values or towards the large?

Summary.   Summarize the statistics for each variate by computing:

- the number of sampling points and the number of valid values;
- the minimum and maximum;
- the mean;
- the median;
- the variance;

- the standard deviation (the square root of the variance);
- the coefficient of variation (optional);
- the skewness (coefficient g_1); and
- the kurtosis (g_2, optional).

A.5 NORMALITY AND TRANSFORMATION

Geostatistical analysis is most efficient when done on variables that have normal, or Gaussian, distributions. Some analyses assume normality. You should therefore examine the form of the distribution of each z.

Symmetric histogram. If the frequency distribution appears symmetric, with a single central peak, try fitting a normal curve to it. If the fit 'looks good', then accept the variate as normal. If not, in what way does it depart from normal? For example, is it flat-topped, or light in the tails? These features may be matched with the coefficient of kurtosis, g_2. A flat-topped distribution suggests that you have more than one population in the image or region—see multiple peaks above.

Skewed histogram. Asymmetry is the most common form of departure from normality, and in particular positive skewness (long upper tail, coefficient $g_1 > 0$). In these circumstances the variance is likely to change from one part of the image or region to another, thereby violating one of the assumptions of stationarity on which analysis is usually based. Consider transforming the recorded z to stabilize the variance. Options are as follows.

- Skew positive, $0 < g_1 < 0.5$. Do not transform.
- Skew positive, $0.5 < g_1 < 1$. Consider transformation to square roots, i.e. $y = \sqrt{z}$.
- Skew positive, $g_1 > 1$. Transform. Try logarithmic transformation first, i.e. $y = \ln z$ or $y = \log_{10} z$. Examine the resultant distribution. If it is approximately normal then accept it. If the result is still skewed then try subtracting a positional constant, a, so that $y = \ln(z - a)$.
  A suitable value for a may be found by fitting the two- and three-parameter lognormal functions to the z.

Other transformations are available if these prove unsatisfactory.

Significance tests for normality are available. *Disregard them when analysing images!* With many pixels you will almost surely discover that the distributions are 'significantly' non-normal. They can be helpful if you have only 100 or so measurements from ground survey.

A.6   SPATIAL DISTRIBUTION

Explore the spatial distribution of each z. Here you might need to treat image data differently from ground data.

Ground data. Make an isarithmic ('contour') map using a reputable program with a well-behaved algorithm for interpolation, such as inverse squared distance weighting or simple bilinear interpolation if the data are dense, and layer shading to indicate the magnitude of z, or $y = f(z)$.

If the data are on a grid then compute the row and column means. Alternatively, find the medians of each row and column.

Is there any trend in them?

Images. If you are analysing images then map the distribution of pixel values using a congenial computer program that will show the individual pixels coloured, or shaded grey, on a scale according to recorded values, or transformed to $y = f(z)$. Compute the row and column means or medians as for gridded ground data.

Examine either kind of map for trends and patches.

Trend. Is there any evident long-range trend over the scene or region?

- If so, what is its form and principal direction?

Patches. Are there patches?

- If so, how big are they on average?
- Are they isotropic?
- If not in which direction are they elongated?

Long-range trend is incompatible with the assumptions of stationarity on which most geostatistical analysis is based. If the trend is strong then consider removing it by some kind of filter, such as a global trend surface, before proceeding further.

Alternatively, adopt a model for z that incorporates the non-stationary trend. This will take you into more advanced technique, and you should consult a specialist about it.

A.7   SPATIAL ANALYSIS: THE VARIOGRAM

The variogram summarizes the spatial distribution of z in the absence of trend. Three variograms are to be distinguished: the experimental variogram; the regional variogram; and the theoretical variogram.

The experimental variogram. This is the variogram that you compute from the data, $z(\mathbf{x}_i)$, $i = 1, 2, \ldots$. For ground data and images on regular rectangular grids compute the semivariances separately along the rows and columns of the grid, incrementing the lag by one sampling interval or pixel at a time, and along the principal diagonals of the grid at intervals of $\sqrt{2}$ sampling intervals.

For irregularly scattered data, and provided you have sufficient (several hundred), compute a variogram in four or more directions by discretizing the lag by both distance and direction. Compute also the variogram ignoring direction, i.e. with lag in distance only.

Plot the results as variance against lag distance with a unique symbol for each direction. Identify the main features, as follows:

Anisotropy. Does the variogram have approximately the same form and values in all directions? If so, then accept it as isotropic and compute an average experimental variogram over all the directions. If not, then in what way do the directions differ?

> **Different spatial scale.** This indicates geometric anisotropy, which might be removed by a simple transformation of the spatial coordinates.

> **Different semivariances.** These indicate 'zonal' anisotropy—there is simply more variance in some directions than in others.

> **Different form.** Look especially for contrasts between convex (decreasing gradient with increasing lag distance) and concave (increasing gradient). This suggests trend in the direction of increasing gradient, and it should be compared with the evidence from the exploratory analysis, above.

Bounds. Does the variogram appear bounded, i.e. does the semivariance reach a maximum within the distance computed or appear as though it would reach a maximum if the lag distance was extended somewhat (bounded)? Alternatively, does it look as though it would increase without limit (unbounded)?

Nugget. Does an imaginary line drawn through the experimental values when projected cut the ordinate at a positive value (not 0)? If so this intercept is known as the nugget variance.

The regional variogram. This is the variogram that you would compute if you had complete information in the region. It is approximated by the experimental variogram.

The theoretical variogram.   This is the variogram of the process that you must imagine generated the field of which the measured data or pixels are a sample.

To proceed further you must fit a mathematical function to the experimental variogram as a model or approximation to the theoretical variogram (see below).

A.8   MODELLING THE VARIOGRAM

1. Match the form of the experimental variogram with those of the common simple valid models for variance in two dimensions. Choose several that appear to have the right form.

2. Fit each of these models in turn using a numerically sound and well-tried program by minimizing a weighted least-squares criterion. Choose weights in proportion to the number of paired comparisons in the experimental values, and set approximate starting values for the non-linear parameters. Tabulate the residual sum of squares and residual mean square as criteria. You may use a more elaborate scheme of weights such as one of those mentioned in Chapter 6 if you wish to model the variogram better near the ordinate.

3. Select the function for which the criteria are least. Plot the fitted function on the same pair of axes as the experimental semivariances. Does the fit appear good on the graph? If not, inspect another. If none appear to fit well, then consider fitting a more complex model by combining two or more simple models from the standard repertoire, and repeat the process.

   In principle, you can always improve the fit of a model by making it more complex, i.e. by increasing the number of parameters in it. To compare functions with different numbers of parameters, calculate the Akaike information criterion (AIC), and choose the model for which the AIC is least. This trades simplicity against goodness of fit. The AIC is defined as

   $$\text{AIC} = -2\ln(\text{maximized likelihood}) + 2 \times (\text{number of parameters}).$$

   For any given experimental variogram it has a variable part:

   $$\hat{A} = n \ln R + 2p,$$

   where n is the number of experimental values, R is the mean squared residual, and p is the number parameters.

   Least-squares fitting minimizes R, but if it is diminished further only by increasing p (n is constant) then there is a penalty, which might be too big.

4. Check that the models that appear to fit accord with prior knowledge. If they do not, then investigate further. You might need to shorten the interval between successive lags, narrow the angular discretization, or extend the maximum distance over which you compute the experimental variogram. You might need to try fitting other models.

5. Tabulate the parameters of the final best model and any others that are almost equally good. You will need the parameters for kriging (below).

A.9   SPATIAL ESTIMATION OR PREDICTION: KRIGING

The aim is to estimate or predict in a spatial sense the values of z at unsampled places, or 'targets', from the data. For images such targets are likely only where there are gaps in a scene. Ordinary kriging smoothes, however, and you might choose to use it to remove short-range noise in the image so that you can see a more general pattern. For ground surveys they are commonplace, and in this section ground survey is assumed. Further, ordinary kriging of z (or $y = \ln z$ for lognormal kriging) is likely to serve in 90% of cases, and only this is covered.

You will need the original data and a legitimate model of the variogram. You now have a number of choices before you.

Punctual or block kriging.   The targets may be points, say $\mathbf{x}_0$, in which case the technique is punctual kriging. Alternatively, they may be small blocks, B, which may be of any reasonable size and shape but are usually square; this is block kriging. The size of block should be determined by the application: what size of block does the user of the predictions want? It should not be determined by the data or the cosmetics of mapping (see below).

Number of data points.   Ordinary kriging computes a weighted average of the data. The weights are determined by the configuration of the data in relation to the target in combination with the variogram model. They do not depend on the measured values, the $z(\mathbf{x}_i)$. Unless the model has a large proportion of nugget variance only the nearest few sampling points carry appreciable weight; more distant points have negligible weight. So kriging is local.

Take the nearest 20 points to the target. If the data points are exceptionally unevenly scattered then take the nearest two or three points in each octant around the target.

Form the kriging equations, and solve them to obtain the weights, the predicted values and the prediction variances (kriging variances).

If you are uncertain how many points to take then experiment with numbers between 4 and 40 and plot their positions in relation to the target and their weights. Do not be alarmed if some weights are negative, provided they are fairly close to 0.

Transformation.   For lognormal kriging the data must transformed to $y = \ln z$ or $y = \log_{10} z$, and the variogram model must be of y. If you want estimates to be of z then you must transform the predicted y back to z.

Kriging for mapping.   Krige at the nodes of a fine square grid. Write the kriged estimates and kriging variances to a file. For an isarithmic display the interval of the grid should be chosen such that it is no more than 2 mm on the final hard copy. The optimality of kriging will not then be noticeably degraded by non-optimal interpolation in the graphics program.

The grid interval need not be related to the block size if you block-krige. The blocks may overlap, or there may be gaps between them.

A.10   MAPPING

Pass the file of kriged estimates and variances to a graphics program for the final display of the results as isarithms or small square cells. Choose colours or grey levels to represent the magnitude of the estimates and variances, as above.

Do not use graphics programs or geographic information systems for geostatistics unless you are in complete control, and you know that they do exactly what you want.

Appendix B
Genstat instructions
for analysis

The analyses summarized above can be done in Genstat using the following commands. The data used as the example are of the exchangeable potassium (K) in the soil of an 80 ha farm (Broom's Barn) in eastern England. The farm was sampled at 40 m intervals on a square grid, and bulked cores of soil to 20 cm were taken and analysed in the laboratory to give 435 values for each variable.

The measured variable (z) is here denoted by z, and the spatial coordinates $(\{x_1, x_2\})$ by x and y in units of 40 m. Unless otherwise defined, variables are vectors, or in the Genstat language, `variates`.

B.1   SUMMARY STATISTICS

Genstat enables you to obtain a statistical summary readily by means of standard functions:

```
calculate zbar=mean(z)
calculate zmed=median(z)
calculate zmax=maximum(z)
calculate zmin=minimum(z)
calculate zmed=median(z)
calculate zvar=var(z)
```

and

```
calculate zsdev=sqrt(zvar)
```

These and other summary statistics can be obtained alternatively using the Genstat procedure `describe`; thus

```
describe [selection=nobs, mean, median, min, max, range, \
    var, sd, skew, kurtosis] z
```

in which `nobs` is the number of non-missing observations and the other options are evident from their names.

B.2   HISTOGRAM

The histogram of z is formed simply by the command

```
histogram [title=!t('Potassium')] z
```

for a device such as a line printer, and by

```
dhistogram [title=!t('Potassium')] z
```

for a high-quality graph. The title within the square brackets is an option.
   You are likely to want to specify the limits to the classes, or 'bins' in statistical jargon. So define a variate containing them:

```
variate [values=10,15...100] binlims
```

Then write

```
dhistogram [limits=binlims; title=!t('Potassium')] z
```

If you want to see what the frequency distribution looks like on the logarithmic scale then z is readily transformed by

```
calculate lz=log10(z)
```

and you can then replace z by lz in the above commands.

B.3   CUMULATIVE DISTRIBUTION

The cumulative distribution of z can be formed by the following set of commands

```
calculate az=sort(z)
calculate cz=cum(az)
calculate nz=nobservations(z)
calculate pz=(!(1...nz)-0.5)/nz
```

in which sort assembles the values in z in order from smallest to largest, and cz contains the accumulated sum.
   To draw a graph of the cumulative distribution you can write the command

```
dgraph x=az; y=pz
```

Further, to show it on a normal probability scale you can convert pz to 'normal equivalent deviates', as follows:

```
calculate nd=ned(pz)
dgraph x=az; y=nd
```

The normal equivalent deviate is the area beneath the standard normal curve of the probability density from $-\infty$ to $G(z)$.

B.4   POSTING

You can plot the data as a posting as follows. You should assemble the outline of the region of interest as pairs of coordinates in two variates, say ox and oy. Then you can write

```
pen 1; linestyle=0; method=point; symbols=4
pen 2; linestyle=1; method=line; symbols=0; join=given
dgraph y=y,ox; x=x,oy; pen=1,2
```

B.5   THE VARIOGRAM

B.5.1   Experimental variogram

You will first want to compute (form) the experimental or sample variogram from your data. The is done using the command fvariogram. It is followed by options and parameters. Below is an example, in which the fvariogram command is preceeded by the declarations of two variates to hold the directions in which you want to compute the variogram and the angles subtended by the segments:

```
variate [nvalues=4] angles; values=!(0,45,90,135)
variate [nvalues=4] segs; values=!(45,45,45,45)
fvariogram [y=y; x=x; step=1; xmax=13; \
  directions=angles; segments=segs] z; \
  variogram=zgam; counts=zcounts; distances=midpts
```

The identifiers zgam, zcounts and midpts are matrices, and you will usually want the results as vectors (variates). These are readily obtained from the matrices by

```
variate vgram [#angles], lag [#angles], count [#angles]
calculate vgram[ ]=zgam$[*;1...4]
calculate lag[ ]=midpts$[*;1...4]
calculate counts[ ]=zcounts$[*;1...4]
```

which you can then print and graph.

An average or omnidirectional variogram can be computed using the fvariogram command, again preceeded by the declarations of two variates to hold the direction and the segments:

```
variate [nvalues=1] angles; values=!(0)
variate [nvalues=1] segs; values=!(180)
fvariogram [y=y; x=x; step=1; xmax=13; \
  directions=angles; segments=segs] z;\
  variogram=zgam; counts=zcounts; distances=midpts
```

Vectors of the identifiers zgam, zcounts and midpts are obtained as follows:

```
variate vgram [#angles], lag [#angles], count [#angles]
```

```
calculate vgram[ ]=zgam$[*;1]
calculate lag[ ]=midpts$[*;1]
calculate counts[ ]=zcounts$[*;1]
```

B.5.2   Fitting a model

Genstat has a procedure, mvariogram, for fitting several standard models to experimental variograms. You can call it, for example, as follows:

```
mvariogram [model=spherical; print=model, summary, estimates; \
    weighting=counts] zgam; counts=zcounts; distances=midpts
```

This will fit a spherical model to the experimental variogram in zgam and midpts with weights proportional to the counts in zcounts. The models available are unbounded linear, bounded linear, circular, spherical, pentaspherical, exponential, Gaussian, Whittle's (besselk1), and power.

The procedure makes use of the fitnonlinear command in Genstat, and you can write models of your own choosing with this command. For example, to compute fit a spherical model you can write:

```
expression spherical
value=!e(c=((1.5*lag/a-0.5*(lag/a)**3*(lag.le.a)+(lag.gt.a))))
model [weights=counts] vrgam
rcycle a; initial=10
fitnonlinear [calculation=spherical] c
```

The expression lag.le.a gives the value 1 if the lag is less than or equal to the range, a, and 0 otherwise. In like manner, lag.gt.a returns 1 if the lag is greater than a and 0 otherwise. These two conditions ensure that the function remains constant once the lag exceeds the range. Notice that only the non-linear part of the model has to be described in the value command. The rcycle command sets an initial value for a. This is to ensure that the search for a solution starts in roughly the right place. Otherwise the program might never converge.

B.6   KRIGING

The kriging facility in Genstat creates a grid of estimates (predictions) for mapping using the command krige, as follows:

```
krige [x=x; y=y; youter=!(1,31); xouter=!(1,18); \
yinner=!(5,20); xinner=!(3,15); block=!(0,0); radius=4.5; \
minpoints=7; maxpoints=20; interval=0.5] \
z; isotropy=isotropic; model=spherical; nugget=0.00476; \
sill=0.01528; range=10.8; predictions=krigest; variances=krigvar
```

B.7 CONTROL

Remember to terminate each Genstat job with

    stop

References

Aitchison, J. and Brown, J. A. C. (1957) *The Lognormal Distribution.* Cambridge University Press, Cambridge.

Akaike, H. (1973) Information theory and an extension of the maximum likelihood principle. In: *Second International Symposium on Information Theory* (eds B. N. Petrov and F. Csáki), pp. 267–281. Akadémiai Kiadó, Budapest.

Armstrong, M. (1998) *Basic Linear Geostatistics.* Springer-Verlag, Berlin.

Atteia, O., Webster, R. and Dubois, J.-P. (1994) Geostatistical analysis of soil contamination in the Swiss Jura. *Environmental Pollution,* **86**, 315–327.

Badr, I., Oliver, M. A., Hendry, G. L. and Durrani, S. A. (1993) Determining the spatial scale of variation in soil radon values using nested sampling and analysis. *Radiation Protection Dosimetry,* **49**, 433–442.

Barnes, R. J. (1989) Sample design for geologic site characterization. In: *Geostatistics,* Volume 2 (ed. M. Armstrong), pp. 809–822. Kluwer Academic Publishers, Dordrecht.

Bartlett, M. S. (1966) *An Introduction to Stochastic Processes,* 2nd edition. Cambridge University Press, Cambridge.

Bierkens, M. F. P. and Burrough, P. A. (1993a) The indicator approach to categorical soil data. I. Theory. *Journal of Soil Science,* **44**, 361–368.

Bierkens, M. F. P. and Burrough, P. A. (1993b) The indicator approach to categorical soil data. II. Application to mapping and land use suitability analysis. *Journal of Soil Science,* **44**, 369–381.

Blackman, R. B. and Tukey, J. W. (1958) *The Measurement of Power Spectra.* Dover Publications, New York.

Brigham, E. O. (1974) *The Fast Fourier Transform.* Prentice Hall, Englewood Cliffs, NJ.

Brus, D. J. and de Gruijter, J. J. (1994) Estimation of non-ergodic variograms and their sampling variance by design-based sampling strategies. *Mathematical Geology,* **26**, 437–454.

Burgess, T. M. and Webster, R. (1980a) Optimal interpolation and isarithmic mapping of soil properties. I. The semi-variogram and punctual kriging. *Journal of Soil Science,* **31**, 315–331.

Burgess, T. M. and Webster, R. (1980b) Optimal interpolation and isarithmic mapping of soil properties. II. Block kriging. *Journal of Soil Science,* **31**, 333–341.

Burgess, T. M. and Webster, R. (1984) Optimal sampling strategy for mapping soil types. I. Distribution of boundary spacings. *Journal of Soil Science,* **35**, 641–654.

Burgess, T. M., Webster, R. and McBratney, A. B. (1981) Optimal interpolation and isarithmic mapping of soil properties. IV. Sampling strategy. *Journal of Soil Science,* **32**, 643–654.

Burrough, P. A., Bregt, A. K., de Heus, M. J. and Kloosterman, E. G. (1985) Complementary use of thermal imagery and spectral analysis of soil properties and wheat yields to reveal cyclic patterns in the Flevopolders. *Journal of Soil Science*, **36**, 141-152.

Chappell, A. and Oliver, M. A. (1997) Analysing soil redistribution in southwest Niger. In: *Geostatics Wollongong '96*, Volume 2 (eds E. Y. Baafi and N. A. Schofield), pp. 961-972. Kluwer Academic Publishers, Dordrecht.

Chauvet, P. (1982) The variogram cloud. In: *Proceedings of the 17th APCOM Symposium* (eds T. B. Johnson and R. J. Barnes), pp. 757-764. Society of Mining Engineers, New York.

Chilès, J.-P. and Delfiner, P. (1999) *Geostatistics: Modeling Spatial Uncertainty*. John Wiley and Sons, New York.

Clark, I., Basinger, K. L. and Harper, W. V. (1989) MUCK—a novel approach to cokriging. In: *Proceedings of the Conference on Geostatistical Sensitivity and Uncertainty Methods for Ground Water Flow and Radionuclide Transport Modeling* (ed. B. E. Buxton), pp 473-493. Battelle Press, Columbus, OH.

Cochran, W. G. (1946) Relative accuracy of systematic and stratified random samples for a certain class of populations. *Annals of Mathematical Statistics*, **17**, 164-177.

Cochran, W. G. (1977) *Sampling Techniques*, 3rd edition. John Wiley and Sons, New York.

Cooley, J. W. and Tukey, J. W. (1965) An algorithm for the machine calculation of complex Fourier series. *Mathematics of Computation*, **19**, 297-301.

Cressie, N. (1985) Fitting variogram models by weighted least squares. *Mathematical Geology*, **17**, 563-586.

Cressie, N. (1990) The origins of kriging. *Mathematical Geology*, **22**, 239-252.

Cressie, N. A. C. (1993) *Statistics for Spatial Data*, revised edition. John Wiley and Sons, New York.

Cressie, N. and Hawkins, D. M. (1980) Robust estimation of the variogram. *Mathematical Geology*, **12**, 115-125.

Dalenius, T., Hájek, J. and Zubrzycki, S. (1961) On plane sampling and related geometrical problems. *Proceedings of the Fourth Berkeley Symposium on Mathematical Statistics and Probability* (ed. J. Neyman), Volume 1, pp. 164-177. University of California Press, Berkeley.

Deutsch, C. V. and Journel, A. G. (1992) *GSLIB Geostatistical Software Library and User's Guide*. Oxford University Press, New York.

Dowd, P. A. (1984) The variogram and kriging: robust and resistant estimators. In: *Geostatistics for Natural Resources Characterization* (eds G. Verly, M. David, A. G. Journel and A. Marechal), pp. 91-106. D. Reidel, Dordrecht.

Durbin, J. and Watson, G. S. (1950) Testing for serial correlation in least squares regression I. *Biometrika*, **37**, 409-428.

Fisher, R. A. (1925) *Statistical Mathods for Research Workers*. Oliver and Boyd, Edinburgh.

Fisher, R. A. and Yates, F. (1963) *Statistical Tables for Biological, Agricultural and Medical Research*, 6th edition. Oliver and Boyd, Edinburgh.

FOEFL (Swiss Federal Office of Environment, Forest and Landscape) (1987) *Commentary on the Ordinance Relating to Pollutants in Soil (VSBo of June 9, 1986)*. FOEFL, Bern.

Frogbrook, Z. L. (1999) The effect of sampling intensity on the reliability of predictions and maps of soil properties. In: *Precision Agriculture '99, Part 1* (ed. J. V. Stafford), pp. 71-80. Sheffield Academic Press, Sheffield.

Frogbrook, Z. L., Oliver, M. A., Salahi, M. and Ellis, R. H. (1999) Comparing the relations in the spatial variation of soil and crop attributes. In: *Precision Agriculture '99, Part 1* (ed. J. V. Stafford), pp. 397-406. Sheffield Academic Press, Sheffield.

Gandin, L. S. (1965) *Objective Analysis of Meteorological Fields.* Israel Program for Scientific Translation, Jerusalem.

Genstat 5 Committee (1993) *Genstat 5 Release 3 Reference Manual.* Oxford University Press, Oxford.

Genstat 5 Committee (1995) *Genstat 5 Release 3.2 Reference Manual Supplement.* Lawes Agricultural Trust, Harpenden.

Genton, M. G. (1998) Highly robust variogram estimation. *Mathematical Geology,* **30**, 213-221.

Gleick, J. (1988) *Chaos.* William Heinemann, London.

Golden Software (1997) *SURFER for Windows, Version 6.* Golden Software, Inc., Golden, CO.

Goovaerts, P. (1994) Performance of indicator algorithms for modelling conditional probability distribution functions. *Mathematical Geology,* **26**, 389-411.

Goovaerts, P. (1997) *Geostatistics for Natural Resources Evaluation.* Oxford University Press, New York.

Goovaerts, P., Webster, R. and Dubois, J.-P. (1997) Assessing the risk of soil contamination in the Swiss Jura using indicator geostatistics. *Environmental and Ecological Statistics,* **4**, 31-48.

Goulard, M. and Voltz, M. (1992) Linear coregionalization model: tools for estimation and choice of cross-variogram matrix. *Mathematical Geology,* **24**, 269-286.

Gower, J. C. (1962) Variance component estimation for unbalanced hierarchical classification. *Biometrics,* **18**, 168-182.

Hallsworth, E. G., Robertson, G. K. and Gibbons, F. R. (1955) Studies in pedogenesis in New South Wales. VII. The 'gilgai' soils. *Journal of Soil Science,* **6**, 1-31.

Hammond, L. C., Pritchet, W. L. and Chew, V. (1958) Soil sampling in relation to soil heterogeneity. *Soil Science Society of America Proceedings,* **22**, 548-552.

Hodge, C. A. H. and Seale, R. S. (1966) *The Soils of the District around Cambridge.* Memoirs of the Soil Survey of Great Britain. Agricultural Research Council, Harpenden.

Isaaks, E. H. and Srivastava, R. M. (1989) *An Introduction to Applied Geostatistics.* Oxford University Press, New York.

Jenkins, G. M. and Watts, D. G. (1968) *Spectral Analysis and its Applications.* Holden-Day, San Francisco.

Journel, A. G. (1983) Non-parametric estimation of spatial distributions. *Mathematical Geology,* **15**, 445-468.

Journel, A. G. (1988) Nonparametric geostatistics for risk and additional sampling assessment. In: *Principles of Environmental Sampling* (ed. L. H. Keith), pp. 45-72. American Chemical Society, Washington, DC.

Journel, A. G. and Huijbregts, C. J. (1978) *Mining Geostatistics.* Academic Press, London.

Jowett, G. H. (1955) Sampling properties of local statistics in stationary stochastic series. *Biometrika,* **42**, 160-169.

Kantey, B. A. and Williams, A. A. B. (1962) The use of soil engineering maps for road projects. *Transactions of the South African Institution of Civil Engineers,* **4**, 149-159.

Kolmogorov, A. N. (1939) Sur l'interpolation et extrapolation des suites stationnaires. *Comptes Rendus de l'Académie des Sciences de Paris,* **208**, 2043-2045.

Kolmogorov, A. N. (1941) Interpolirovanie i ekstrapolirovanie statsionarnykh sluchainykh posledovatel' nostei (Interpolated and extrapolated stationary random sequences). *Izvestya Akademiya Nauk SSSR, Seriya Matematicheskaya*, 5(1).

Krige, D. G. (1966) Two-dimensional weighted moving average trend surfaces for ore-evaluation. *Journal of the South African Institute of Mining and Metallurgy*, 66, 13-38.

Krumbein, W. C. and Slack, H. A. (1956) Statistical analysis of low-level radioactivity of Pennsylvanian black fissile shale in Illinois. *Bulletin of the Geological Society of America*, 67, 739-762.

Lake, J. V., Bock, G. R. and Goode, J. A. (1997) *Precision Agriculture: Spatial and Temporal Variability of Environmental Quality*. John Wiley and Sons, Chichester.

Langsaetter, A. (1926) Om beregning av middelfeilen ved regelmessige linjetakseringer. *Meddelanden fra det norske Skogsforsøksvesen*, 2 h 7, 5-47.

Lark, R. M. (2000) A comparison of some robust estimators of the variogram for use in soil survey. *European Journal of Soil Science*, 51, 137-157.

Laslett, G. M., McBratney, A. B., Pahl, P. J. and Hutchinson, M. F. (1987) Comparison of several spatial prediction methods for soil pH. *Journal of Soil Science*, 38, 325-341.

Leenaers, H., Okx, J. P. and Burrough, P. A. (1990) Employing elevation data for efficient mapping of soil pollution on flood plains. *Soil Use and Management*, 6, 105-114.

Leenhardt, D., Voltz, M., Bornand, M. and Webster, R. (1994) Evaluating soil maps for prediction of soil water properties. *European Journal of Soil Science*, 45, 293-301.

Lemmer, I. C. (1984) Estimating local recoverable reserves via indicator kriging. In: *Geostatistics for Natural Resources Characterization* (eds G. Verly, M. David, A. G. Journel and A. Marechal), pp. 349-364. D. Reidel, Dordrecht.

Marcuse, S. (1949) Optimum allocation and variance components in nested sampling with application to chemical analysis. *Biometrics*, 5, 189-206.

Mardia, K. V. and Jupp, P. F. (2000) *Directional Statistics*. John Wiley and Sons, Chichester.

Maréchal, A. (1976) The practice of transfer functions: Numerical methods and their application. In: *Advanced Geostatistics in the Mining Industry* (eds M. Guarascio, M. David and C. Huijbregts), pp. 253-276. D. Reidel, Dordrecht.

Marquardt, D. W. (1963) An algorithm for least-squares estimation of nonlinear parameters. *Journal of the Society of Industrial and Applied Mathematics*, 11, 431-441.

Matérn, B. (1960) Spatial variation: Stochastic models and their applications to problems in forest surveys and other sampling investigations. *Meddelanden från Statens Skogforskningsinstitut*, 49, 1-144.

Matheron, G. (1963) Principles of geostatistics. *Economic Geology*, 58, 1246-1266.

Matheron, G. (1965) *Les variables régionalisées et leur estimation*. Masson, Paris.

Matheron, G. (1969) *Le krigeage universel*. Cahiers du Centre de Morphologie Mathématique, École des Mines de Paris, Fontainebleau.

Matheron, G. (1973) The intrinsic random functions and their applications. *Advances in Applied Probability*, 5, 439-468.

Matheron, G. (1976) A simple substitute for conditional expectation: The disjunctive kriging. In: *Advanced Geostatistics in the Mining Industry* (eds M. Guarascio, M. David and C. Huijbregts), pp 221-236. D. Reidel, Dordrecht.

Matheron, G. (1979) *Recherche de simplification dans un problème de cokrigeage*. Publication N-628. Centre de Géostatistique, École des Mines de Paris, Fontainebleau.

Matheron, G. (1989) *Estimating and Choosing*. Springer-Verlag, Berlin.

McBratney, A. B. and Webster, R. (1981) Detection of ridge and furrow patterns by spectral analysis of crop yield. *International Statistical Review*, **49**, 45–52.

McBratney, A. B. and Webster, R. (1983) Optimal interpolation and isarithmic mapping of soil properties. V. Coregionalization and multiple sampling strategy. *Journal of Soil Science*, **34**, 137–162.

McBratney, A. B. and Webster, R. (1986) Choosing functions for semivariograms of soil properties and fitting them to sampling estimates. *Journal of Soil Science*, **37**, 617–639.

McBratney, A. B., Webster, R. and Burgess, T. M. (1981) The design of optimal sampling schemes for local estimation and mapping of regionalized variables. *Computers and Geosciences*, **7**, 331–334.

McBratney, A. B., Webster, R., McLaren, R. G. and Spiers, R. B. (1982) Regional variation of extractable copper and cobalt in the topsoil of south-east Scotland. *Agronomie*, **2**, 969–982.

McCullagh, M. J. (1976) Estimation by kriging of the reliablity of the Trent telemetry network. *Computer Applications*, **2**, 357–374.

Mercer, W. B. and Hall, A. D. (1911) Experimental error of field trials. *Journal of Agriculture Science, Cambridge*, **4**, 107–132.

Miesch, A. T. (1975) Variograms and variance components in geochemistry and ore evaluation. *Geological Society of America Memoir*, **142**, 333–340.

Ministry of Agriculture, Fisheries and Food (1986) *The Analysis of Agricultural Materials*, 3rd edition. MAFF Reference Book 427. Her Majesty's Stationery Office, London.

Moffat, A. J., Catt, J. A., Webster, R. and Brown, E. H. (1986) A re-examination of the evidence for a Plio-Pleistocene marine transgression on the Chiltern Hills. I. Structures and surfaces. *Earth Surface Processes and Landforms*, **11**, 95–106.

Morse, R. K. and Thornburn, T. H. (1961) Reliability of soil units. In: *Proceedings of the 5th International Conference on Soil Mechanics and Foundation Engineering, Volume 1*, pp. 259–262. Dunod, Paris.

Mulla, D. J. (1997) Geostatistics, remote sensing and precision farming. In: *Precision Agriculture: Spatial and Temporal Variability of Environmental Quality* (eds J. V. Lake, G. R. Bock and J. A. Goode), pp. 100–115. John Wiley and Sons, Chichester.

Muñoz-Pardo, J. F. (1987) Approche géostatistique de la variabilité spatiale des milieux géophysique. Thèse de Docteur-Ingénieur, Université de Grenoble et l'Institut National Polytechnique de Grenoble.

Myers, D. E. (1982) Matrix formulation of cokriging. *Mathematical Geology*, **14**, 249–257.

Myers, D. E. (1991) Pseudo-cross-variograms, positive definiteness, and cokriging. *Mathematical Geology*, **23**, 805–816.

Olea, R. A. (1975) *Optimum Mapping Techniques Using Regionalized Variable Theory*. Series on Spatial Analysis, no. 2. Kansas Geological Survey, Lawrence.

Olea, R. A. (1999) *Geostatistics for Engineers and Earth Scientists*. Kluwer Academic Publishers, Boston.

Oliver, M. A. and Badr, I. (1995) Determining the spatial scale of variation in soil radon concentration. *Mathematical Geology*, **27**, 893–922.

Oliver, M. A. and Webster, R. (1986) Combining nested and linear sampling for determining the scale and form of spatial variation of regionalized variables. *Geographical Analysis*, **18**, 227–242.

Oliver, M. A. and Webster, R. (1987) The elucidation of soil pattern in the Wyre Forest of the West Midlands, England. II. Spatial distribution. *Journal of Soil Science*, **38**, 293–307.

Oliver, M. A., Webster, R. and McGrath, S. P. (1996) Disjunctive kriging for environmental management. *Environmentrics*, **7**, 333-358.

Oliver, M. A., Webster, R. Edwards, K. J. and Whittington, G. (1997) Multivariate, autocorrelation and spectral analyses of a pollen profile from Scotland and evidence of periodicity. *Review of Palaeobotany and Palynology*, **96**, 121-141.

Oliver, M. A. and Webster, R. and Slocum, K. (2000) Filtering SPOT imagery by kriging analysis. *International Journal of Remote Sensing*, **21**, 735-752.

Omre, H. (1987) Bayesian kriging—merging observations and qualified guesses in kriging. *Mathematical Geology*, **19**, 25-39.

Pannatier, Y. (1995) *Variowin. Software for Spatial Analysis in 2D.* Springer-Verlag, New York.

Papritz, A. and Webster, R. (1995a) Estimating temporal change in soil monitoring: I. Statistical theory. *European Journal of Soil Science*, **46**, 1-12.

Papritz, A. and Webster, R. (1995b) Estimating temporal change in soil monitoring: II. Sampling from simulated fields. *European Journal of Soil Science*, **46**, 13-27.

Papritz, A., Künsch, H. R. and Webster, R. (1993) On the pseudo cross-variogram. *Mathematical Geology*, **25**, 1015-1026.

Parzen, E. (1961) Mathematical considerations in the estimation of spectra. *Technometrics*, **3**, 167-190.

Press, W. H., Flannery, B. P., Teukolsky, S. A. and Vetterling, W. T. (1992) *Numerical Recipes in Fortran*, second edition. Cambridge University Press, Cambridge.

Priestley, M. B. (1981) *Spectral Analysis and Time Series.* Academic Press, London.

Quenouille, M. H. (1949) Problems in plane sampling. *Annals of Mathematical Statistics*, **20**, 355-375.

Ratkowsky, D. A. (1983) *Nonlinear Regression Modeling.* Marcel Dekker, New York.

Rendu, J.-M. (1980) Disjunctive kriging: Comparison of theory with actual results. *Mathematical Geology*, **12**, 305-320.

Rivoirard, J. (1994) *Introduction to Disjunctive Kriging and Non-linear Geostatistics.* Oxford University Press, Oxford.

Ross, G. J. S. (1987) *Maximum Likelihood Program.* Numerical Algorithms Group, Oxford.

SAS Institute (1985) *SAS User's Guide: Statistics, Version 5 Edition.* SAS Institute Inc., Cary, NC.

Scott, R. M., Webster, R. and Lawrance, C. J. (1971) *A Land System Atlas of Western Kenya.* Military Vehicles and Engineering Establishment, Christchurch, Dorset.

Shafer, J. M. and Varljen, M. D. (1990) Approximation of confidence limits on sample semivariograms from single realizations of spatially correlated random fields. *Water Resources Research*, **26**, 1787-1802.

Shepard, D. (1968) A two-dimensional interpolation function for irregularly-spaced data. *Proceedings of the Association for Computing Machinery (1968)*, 517-523.

Sibson, R. (1981) A brief description of natural neighbour interpolation. In: *Interpreting Multivariate Data* (ed. V. Barnett), pp. 21-36. John Wiley and Sons, Chichester.

Snedecor, G. W. and Cochran, W. G. (1967) *Statistical Methods*, 6th edition, Iowa State University Press, Ames.

Sullivan, J. (1984) Conditional recovery estimation through probability kriging: Theory and practice. In: *Geostatistics for Natural Resources Characterization* (eds G. Verly, M. David, A. G. Journel and A. Marechal), pp. 365-384. D. Reidel, Dordrecht.

Taylor, C. C. and Burrough, P. A. (1986) Multiscale sources of spatial variation in soil. III. Improved methods for fitting the nested model to one-dimensional semi-variograms. *Mathematical Geology*, **18**, 811-821.

Tukey, J. W. (1977) *Exploratory Data Analysis.* Addison-Wesley, Reading, MA.

Vauclin, M., Vieira, S. R., Vachaud, G. and Nielsen, D. R. (1983) The use of co-kriging with limited field data. *Soil Science Society of America Journal*, **47**, 175–184.

Von Neumann, J. (1941) Distribution of the ratio of the mean square difference to the variance. *Annals of Mathematical Statistics*, **12**, 367–395.

Von Steiger, B., Webster, R., Schulin, R. and Lehmann, R. (1996) Mapping heavy metals in polluted soil by disjunctive kriging. *Environmental Pollution*, **94**, 205–215.

Wackernagel, H. (1994) Cokriging versus kriging in regionalized multivariate data analysis. *Geoderma*, **62**, 83–92.

Wackernagel, H. (1995) *Multivariate Geostatistics*. Springer-Verlag, Berlin.

Webster, R. (1977) Spectral analysis of gilgai soil. *Australian Journal of Soil Research*, **15**, 191–204.

Webster, R. (1991) Local disjunctive kriging of soil properties with change of support. *Journal of Soil Science*, **42**, 301–318.

Webster, R. (1994) Estimating trace elements in soil: A case study in cobalt deficiency. In: *Introduction to Disjunctive Kriging and Non-linear Geostatistics* (ed. J. Rivoirard), pp. 128–145. Oxford University Press, Oxford.

Webster, R. and Beckett, P. H. T. (1968) Quality and usefulness of soil maps. *Nature (London)*, **219**, 680–682.

Webster, R. and Beckett, P. H. T. (1970) Terrain classification and evaluation using air photography: a review of recent work at Oxford. *Photogrammetria.* **26**, 51–75.

Webster, R. and Boag, B. (1992) A geostatistical analysis of cyst nematodes in soil. *Journal of Soil Science*, **43**, 583–595.

Webster, R. and Burgess, T. M. (1980) Optimum interpolation and isarithmic mapping of soil properties. III. Changing drift and universal kriging. *Journal of Soil Science*, **31**, 505–524.

Webster, R. and Butler, B. E. (1976) Soil survey and classification studies at Ginninderra. *Australian Journal of Soil Research*, **14**, 1–26.

Webster, R. and McBratney, A. B. (1987) Mapping soil fertility at Broom's Barn by simple kriging. *Journal of the Science of Food and Agriculture*, **38**, 97–115.

Webster, R. and McBratney, A. B. (1989) On the Akaike information criterion for choosing models for variograms of soil properties. *Journal of Soil Science*, **40**, 493–496.

Webster, R. and Oliver, M. A. (1989) Optimal interpolation and isarithmic mapping of soil properties. VI. Disjunctive kriging and mapping the conditional probability. *Journal of Soil Science*, **40**, 497–512.

Webster, R. and Oliver, M. A. (1990) *Statistical Methods in Soil and Land Resource Survey*. Oxford University Press, Oxford.

Webster, R. and Oliver, M. A. (1992) Sample adequately to estimate variograms of soil properties. *Journal of Soil Science*, **43**, 177–192.

Webster, R. and Oliver, M. A. (1997) Software review. *European Journal of Soil Science*, **43**, 173–175.

Webster, R. and Rivoirard, J. (1991) Copper and cobalt deficiency in soil: A study using disjunctive kriging. In: *Cahiers de Géostatistique, Compte-Rendu des Journées de Géostatistique*, Volume 1, pp. 205–223. École des Mines de Paris, Fontainebleau.

Webster, R., Atteia, O. and Dubois, J.-P. (1994) Coregionalization of trace metals in the soil in the Swiss Jura. *European Journal of Soil Science*, **45**, 205–218.

Whittle, P. (1954) On stationary processes in the plane. *Biometrika*, **41**, 434–449.

Wiener, N. (1949) *Extrapolation, Interpolation and Smoothing of Stationary Time Series*. MIT Press, Cambridge, MA.

Wold, H. (1938) *A Study in the Analysis of Stationary Time Series*. Almqvist and Wiksell, Uppsala.

Wood, G., Oliver, M. A. and Webster, R. (1990) Estimating soil salinity by disjunctive kriging. *Soil Use and Management*, **6**, 97-104.

Yaglom, A. M. (1987) *Correlation Theory of Stationary and Related Random Functions. Volume 1: Basic Results.* Springer-Verlag, New York.

Yates, F. (1948) Systematic sampling. *Philosophical Transactions of the Royal Society of London A*, **241**, 345-377.

Yates, F. (1981) *Sampling Methods for Censuses and Surveys*, 4th edition. Griffin, London.

Yates, S. R. and Yates, M. V. (1988) Disjunctive kriging as an approach to management decision making. *Soil Science Society of America Journal*, **52**, 1554-1558.

Yates, S. R., Warrick, A. W. and Myers, D. E. (1986a) Disjunctive kriging. I. Overview of estimation and conditional probability. *Water Resources Research*, **22**, 615-621.

Yates, S. R., Warrick, A. W. and Myers, D. E. (1986b) Disjunctive kriging. II. Examples. *Water Resources Research*, **22**, 623-630.

Yfantis, E. A., Flatman, G. T. and Behar, J. V. (1987) Efficiency of kriging estimation for square, triangular and hexagonal grids. *Mathematical Geology*, **19**, 183-205.

Youden, W. J. and Mehlich, A. (1937) Selection of efficient methods for soil sampling. *Contributions of the Boyce Thompson Institute for Plant Research*, **9**, 59-70.

Yule, G. U. and Kendall, M. G. (1950) *An Introduction to the Theory of Statistics*, 14th edition. Griffin, London.

Index

CHESTER COLLEGE LIBRARY